LEVERAGING THE PRIVATE SECTOR

Management-Based Strategies for Improving Environmental Performance

EDITED BY

CARY COGLIANESE AND JENNIFER NASH

RESOURCES FOR THE FUTURE
WASHINGTON, DC, USA

An RFF Press book
Published by Resources for the Future
1616 P Street NW
Washington, DC 20036–1400
USA
www.rffpress.org

Library of Congress Cataloging-in-Publication Data

Leveraging the private sector : management-based strategies for improving environ-mental performance / [edited by] Cary Coglianese and Jennifer Nash.
    p. cm.
  Includes bibliographical references and index.
  ISBN 1-891853-95-3 (cloth : alk. paper) — ISBN 1-891853-96-1 (pbk. : alk. paper)
  1. Industrial management—Environmental aspects—United States. 2. Environmen-tal policy—United States. 3. Environmental protection—Standards—United States. 4. Environmental management—United States—Evaluation.
I. Coglianese, Cary. II. Nash, Jennifer, 1956-
HD30.255.L48 2006
658.4'083--dc22   2006000042

The paper in this book meets the guidelines for permanence and durability of the Com-mittee on Production Guidelines for Book Longevity of the Council on Library Resources. This book was typeset by Peter Lindeman. It was copyedited by Hope Steele. The cover was designed by Circle Graphics.

ISBN 1-891853-95-3 (cloth)   ISBN 1-891853-96-1 (paper)

About Resources for the Future *and* RFF Press

RESOURCES FOR THE FUTURE (RFF) improves environmental and natural resource policymaking worldwide through independent social science research of the highest caliber. Founded in 1952, RFF pioneered the application of economics as a tool for developing more effective policy about the use and conservation of natural resources. Its scholars continue to employ social science methods to analyze critical issues concerning pollution control, energy policy, land and water use, hazardous waste, climate change, biodiversity, and the environmental challenges of developing countries.

RFF PRESS supports the mission of RFF by publishing book-length works that present a broad range of approaches to the study of natural resources and the environment. Its authors and editors include RFF staff, researchers from the larger academic and policy communities, and journalists. Audiences for publications by RFF Press include all of the participants in the policymaking process—scholars, the media, advocacy groups, NGOs, professionals in business and government, and the public.

Contents

Foreword

Since at least the publication of Garrett Hardin's paradigm-setting article "The Tragedy of the Commons" in 1968, theories about protecting the environment in the industrialized world have revolved mainly around two stereotypes, which may be called *regulatory environmentalism* and *free-market environmentalism*. Regulatory environmentalism is the dominant paradigm, according to which so-called market failures lead to environmental degradation that must be redressed by governmental regulation. The second, less-prominent model of free-market environmentalism—also discussed by Hardin and later developed by others—posits that environmental problems result from failures to specify private property rights properly. Its adherents argue that the best path to protecting the environment is not bureaucratic control but an improvement in the system of market incentives. Although regulatory environmentalists and free-market environmentalists often stridently disagree about the solutions to environmental problems, they share a common image of industrial firms as automatons responding in a self-interested manner to the external incentives with which they are presented either by government or by the market.

These incentives, whether they take the form of command-and-control regulations or emissions fees and tradable permits, encourage firms to do one of two things: either reduce emissions outputs or install pollution-control technologies. They tend to treat the industrial firm itself as a black box and view environmental problems as merely the failure of proper incentives from outside the firm to affect its inputs and outputs. Yet the work in this book, as well as much of the other path-breaking work by Cary Coglianese and Jennifer Nash, shows that there is a critical link that cannot be ignored between inputs and outputs. That link is the management that takes place inside the firm. The style and commitment of a firm's management is an important factor that must be taken into account in explaining observed variations in corporate behavior and environmental performance. Any adequate solution to environmental problems will depend in part on affecting that management.

My own experience across government, academia, and the private sector is that every company I have ever worked with that I would consider a superior environmental performer has one or more individual managers who are committed to protecting the environment. Management does matter. As we know, organizational economics has long maintained that corporations are not simple profit-maximizers, but executives have some leeway at the margins to pursue their own goals and objectives within the constraints of achieving satisfactory economic returns. Some managers possibly use their power as imperfectly supervised agents of the stockholders to pursue personal values, such as protecting the environment, at least at the margins. Some stockholders may also approve of some corporate activities to protect and enhance the environment, even where "green investing" is not their primary or stated goal.

There is no question that these behaviors are more prominent in business enterprises where external incentives from customers favor a "more environmentally correct than thou" strategy. But the fact remains, as documented in the pages of this book, that not all those engaged in the same lines of business are equally effective at protecting the environment. The internal management factors that help to explain the distribution of behaviors are as worthy of academic and public policy attention as are the external incentives aimed at inputs and outputs.

Recent scholarship has argued that the selfish, unenlightened model of human nature posited by Hardin and almost everyone after him is too simple, and that sometimes human beings do act cooperatively to protect the environment. An interesting and important question that this book raises is, under what circumstances private efforts can supplement those of government and the market to protect and enhance the environment. Can we find a way to shape not only the inputs and outputs of firms, but also the behavior and commitments of the individual managers and the organization settings in which they work?

To be sure, no one believes that private efforts alone will suffice to protect the environment. James Madison reminds us in *The Federalist 51* that governments are necessary because people are not angels; for the same reason, governmental regulation and economic incentives will remain necessary to protect the environment. But clearly, private sector management is, and must be, the front line in any effort to address environmental problems. Finding ways to leverage the private sector will be vital for the next generation of environmental protection. Government, market incentives, and private efforts must all be forged together into a composite strategy to elicit appropriate management responses that will protect the environment. The challenge ahead is to choose the most successful mix; to do that, we must understand the preconditions and limitations of each of the various approaches.

This wonderful book begins to address this core challenge by making visible an approach to protecting the environment that most of us have overlooked. The analysis provided here deserves to be studied both by scholars and policy actors who seek to develop better responses to environmental problems, as well as by the private sector leaders who will be instrumental to the success of any future effort to protect the environment.

E. DONALD ELLIOTT*

---

*Professor (adjunct) of Law, Yale Law School and Georgetown University Law Center; Partner, Willkie Farr & Gallagher LLP, Washington, DC. Professor Elliott was General Counsel of EPA during the George H.W. Bush administration, 1989-1991.

Acknowledgments

In this book, we address a set of questions that follows from the findings of our previous book, *Regulating from the Inside: Can Environmental Management Systems Achieve Policy Goals?* That book showed that private sector firms using environmental management systems can prevent pollution and better comply with environmental regulations. If environmental management systems are associated with improved environmental performance, the question arises of whether governmental and private institutions should require or at least encourage firms to adopt these systems. That question motivated still larger ones to which this book is devoted: How can actors outside businesses—such as regulatory agencies, trade associations, and environmental and community groups—shape the style and effectiveness of the environmental management that takes place inside the firm? What are the most effective ways of leveraging private sector management to achieve environmental gains? What advantages and disadvantages do such new, management-based strategies possess compared with more traditional strategies?

To address these questions, we have commissioned the papers that make up the chapters in this book. Authored by leading experts in law, economics, political science, and public policy, these studies show both the potential and the limits of management-based strategies. Taken together, their empirical findings indicate that the quality and style of a firm's management can affect its environmental performance, that outside forces can shape firms' internal management, and that at least sometimes these outside efforts to leverage management can result in measurable environmental gains. Even though some of the studies in this book also show that management-based strategies are no panacea, we believe this book does demonstrate that they are distinctive and worth analyzing.

This book came about only as the result of many people's contributions. Initial drafts of the chapters provided the basis for a productive discussion held at the headquarters of Resources for the Future in Washington, DC, in 2003. That meeting was organized by the Regulatory Policy Program at the John F. Kennedy School of Government at Harvard University and cosponsored by the Charles G. Koch Charitable Foundation, the U.S. Environmental Protection Agency Office of Policy, Economics, and Innovation (Grant No. R-83056701), the American Chemistry Council, Resources for the Future/RFF Press, the Multi-State Working Group Policy Academy, and the Mossavar-Rahmani Center for Business and Government at the John F. Kennedy School of Government. Additional support came from the Public Service Enterprise Group and from Sidley Austin Brown & Wood LLP.

We benefited enormously from the comments of all those who participated in the Washington, DC, conference. In addition to the assemblage of authors who wrote the chapters in this book, we are indebted to the following people who played key roles at the conference and generously offered their insights: Max Bazerman, Jay Benforado, Leslie Carothers, James Connaughton, Terry Davies, Donald Elliott, Winston Hickox, Jeffrey Holmstead, Rachel Jean-Baptiste, Debra Knopman, Mindy Lubber, Tom Lyon, Jason Morrison, Paul Portney, Susan Rose-Ackerman, Lynn Scarlett, Robert Stephens, Eric Svenson, Barton Thompson, John Walke, and Terry Yosie. Of course, the views expressed in this book do not necessarily represent the opinions of these individuals or of any of the sponsoring organizations.

We are grateful to Ben Gerber for research assistance, Bernard Cahill for logistical support, and Hope Steele for invaluable editorial guidance. We benefited also from the comments of our anonymous reviewers as well as the staff at RFF Press, particularly Grace Hill and Meg Keller. Finally, we are extremely thankful to Don Reisman, the director of RFF Press, for his more than abundant encouragement and patience in seeing this project through from start to finish.

CARY COGLIANESE
JENNIFER NASH

Contributors

Richard N. L. (Pete) Andrews is the Thomas Willis Lambeth Distinguished Professor in the Department of Public Policy at the University of North Carolina at Chapel Hill, with joint appointments in city and regional planning, environmental sciences and engineering, and the Carolina Environmental Program. He has published widely on the history, instruments, and analysis of environmental policy, including *Environmental Policy and Administrative Change: Implementation of the National Environmental Policy Act* and *Managing the Environment, Managing Ourselves: A History of American Environmental Policy*.

Lori Snyder Bennear is an assistant professor of environmental economics and policy at the Nicholas School of the Environment and Earth Sciences at Duke University. Her research focuses on estimating the effect of different regulatory innovations on measures of facility-level environmental performance, such as pollution levels, chemical use, and technology choice. She has published in *Environment* magazine, the *Journal of Regulatory Economics*, and the *American Economic Review*.

Cary Coglianese is the Edward B. Shils Professor of Law and professor of political science at the University of Pennsylvania, where he is the director of the Penn Program on Regulation. He is also a senior research fellow at the Mossavar-Rahmani Center for Business and Government at Harvard University's John F. Kennedy School of Government, where he served on the faculty for twelve years. Coglianese has taught regulatory policy as a visiting professor at the Stanford Law School and the Vanderbilt University Law School, and is currently a vice-chair of the Committee on Innovation, Management Systems, and Trading of the American Bar Association's environmental law section. His research focuses on a variety of issues of regulation and administrative law, with an emphasis on the evaluation of alternative and innovative regulatory strategies. The author of numerous articles in leading academic journals, he also coedited *Regulating from the Inside: Can Environmental Management Systems Achieve Policy Goals?*, published by RFF Press.

Daniel ("David") Edwards, Jr. is a 2004 Master of City and Regional Planning graduate of the University of North Carolina at Chapel Hill. He worked for three years on the National Database on Environmental Management Systems, a longitudinal research study sponsored by the U.S. Environmental Protection Agency on implementation of ISO 14001 environmental management systems, and subsequently for three years on the EPA-funded survey project reported here.

Andrew M. Hutson is a research associate and doctoral candidate in the Department of Public Policy at the University of North Carolina at Chapel Hill. His research areas include corporate environmental management and the influence of global value chain relationships on social and environmental behaviors, particularly in industrializing societies. Before returning to graduate school, he was a business process analyst with Accenture and spent time as a volunteer with the RARE Center for Tropical Conservation in Honduras.

Jason Scott Johnston is the Robert G. Fuller, Jr. Professor of Law and director of the Program on Law and the Environment (POLE) at the University of Pennsylvania Law School. His current work focuses on the impact of tradable permit regimes on technology choice, the rule of capture in natural resource law, the positive theory of regulatory federalization, and optimal public pricing for excludable natural resource use. His articles have appeared in the *Yale Law Journal*, the *Journal of Law, Economics and Organization*, and other leading academic journals.

Robert A. Kagan, professor of political science and law at the University of California, Berkeley, has conducted empirical research on comparative legal and regulatory systems, law enforcement by administrative agencies, and business response to regulation. His publications include *Regulatory Justice; Going by the Book* (with Eugene Bardach); *Regulatory Encounters* (with Lee Axelrad); *Adversarial Legalism: The American Way of Law;* and *Shades of Green* (with Neil Gunningham and Dorothy Thornton).

Andrew A. King is an associate professor of strategy at the Tuck School of Business at Dartmouth College. His principal research areas include market entry and location strategies, environmental strategy, and industry self-regulation. He holds a PhD in management from the Massachusetts Institute of Technology and mechanical engineering degrees from the University of California at Berkeley and Brown University.

Paul R. Kleindorfer is the Anheuser Busch Professor of Management Science and professor of business and public policy at the Wharton School of the University of Pennsylvania. His research has focused on risk management and regulation in energy and environment, and his recent books include *Catastrophe Insurance: Supply, Demand and Regulation*.

Howard Kunreuther is the Cecilia Yen Koo Professor of Decision Sciences and Public Policy at the Wharton School, University of Pennsylvania as well as the codirector of the Wharton Risk Management and Decision Processes Center. He is a fellow of the American Association for the Advancement of Science (AAAS) and distinguished fellow of the Society for Risk Analysis, receiving the Society's Distinguished Achievement Award in 2001. He is the author of *Catastrophe Modeling: A New Approach to Managing Risk* (with Patricia Grossi) and *Wharton on Making Decisions* (with Stephen Hoch).

Shelley H. Metzenbaum is a visiting professor at the University of Maryland School of Public Policy, where she runs an executive education program on regulatory policy for federal science agencies. She also leads the Environmental Compliance Consortium to improve the measurement and management of government environmental programs, and consults and writes on performance measurement and management. Metzenbaum previously served as associate administrator for Regional Operations and State/Local Relations at EPA, undersecretary of Massachusetts Environmental Affairs, and director of the Massachusetts Capital Budget.

Jennifer Nash is executive director of the Corporate Social Responsibility Initiative at the John F. Kennedy School of Government at Harvard University and director of the Regulatory Policy Program at the Kennedy School's Mossavar-Rahmani Center for Business and Government. She conducts empirical research on government regulation, industry self-regulation, and international codes of conduct and has published in journals such as *Administrative Law Review, Business, Strategy and the Environment, Environment, Annual Review of Energy and the Environment,* and *California Management Review.* She co-edited *Regulating from the Inside: Can Environmental Management Systems Achieve Policy Goals?,* also published by RFF Press. She is a member of EPA's National Advisory Council for Environmental Policy and Technology.

Tapas K. Ray is an economist at the National Institute of Occupational Safety and Health. He joined the Centers for Disease Control and Prevention as a Prevention Effectiveness Fellow in 2004 after completing his PhD in economics at the University of Connecticut. His areas of specialization are industrial organization and environmental and regulatory economics. Involved in projects of national and international significance that relate to workers' safety and health, his current interests are work stress, long work hours, and making business cases for ergonomic interventions.

Peter Schmeidler is a senior fellow at the Risk Management and Decision Processes Center at the Wharton School of the University of Pennsylvania, where he works on projects covering the use of third parties to evaluate a facility's environmental, safety, and health management systems. He is editor of the *Risk Management Review*, the Risk Center newsletter. He retired from the Rohm and Haas Company after 40 years of service in the area of process design, development, safety, and technology.

Kathleen Segerson is professor in the Department of Economics at the University of Connecticut. Her research focuses on the incentive effects of alternative environmental policy instruments. She has served on several committees of the EPA Science Advisory Board and the National Research Council, and has published numerous papers in journals such as the *Journal of Environmental Economics and Management*, the *American Journal of Agricultural Economics*, and the *Journal of Legal Studies*.

Part I: Management-Based Strategies

1

Management-Based Strategies

An Emerging Approach to Environmental Protection

Cary Coglianese and Jennifer Nash

For decades, policymakers in the United States have required firms either to install specific pollution-control technologies or to ensure that emissions from their smoke stacks or drain pipes stay within specific limits. This two-pronged approach of environmental technology requirements and emissions limits has succeeded in reducing certain types of environmental problems. Yet this approach, which forms the backbone of the nation's environmental protection system, has for the most part ignored what goes on inside the firms and facilities that actually generate pollution. As long as companies deployed the appropriate technologies or met the specified emissions targets, how they managed their facilities and what systems they used to monitor their environmental performance mattered to no one other than the company managers themselves, and perhaps their corporate boards and shareholders. Policymakers and others interested in environmental protection treated the firm itself as a black box (Salzman and Thompson 2003, *182*).

The tendency to ignore what goes on inside the firm is beginning to change. As environmental problems persist, even after decades of conventional regulation, policymakers, trade associations, and nongovernmental organizations are searching for alternative solutions. Among the most innovative alternatives are strategies that focus directly on how companies manage their internal operations. Previously overlooked in the literature on environmental protection, these management-based strategies are distinctive in that they seek to penetrate and shape what goes on inside private sector firms. What had previously been treated as a black box—the firm itself and its internal management—is now being exposed as a direct and explicit target of government regulators and others who seek to induce firms to improve their environmental performance.

For example, even though the extensive amendments Congress made to the Clean Air Act in 1990 contained numerous conventional regulatory provisions, they also included a new regulatory program that targeted the internal management of large chemical facilities in an attempt to reduce the risk of catastrophic accidents. During the 1980s, U.S. chemical facilities had experienced over a dozen accidents that could have potentially resulted in disastrous outcomes like those in Bhopal, India, in 1984, where thousands of people died (*Congressional Record* 1989). In response, members of Congress sought to improve what they perceived as a critical variable contributing to these accidents—internal management. Under Section 112(r) of the revised Act, chemical facilities managers must now undertake a three-step management process. They must (1) inventory the amounts and types of hazardous chemicals used, (2) develop scenarios about potential catastrophic accidents, and (3) write and implement an accident prevention plan.[1]

Government policymakers have not been the only ones to target firms' internal management in an attempt to improve private sector environmental performance. Around the same time that Congress deliberated over amending the Clean Air Act, representatives of the Chemical Manufacturers Association, a major industry trade association, met across the Potomac River in Northern Virginia to develop their own private sector solution to the problem of chemical accidents (Lodge and Rayport 1991).[2] The result of their efforts, known as the Responsible Care initiative, has much in common with what Congress eventually mandated in Section 112(r) of the Clean Air Act. Responsible Care requires all trade association members to identify those aspects of their operations that could cause chemical accidents and to work with employees and members of local communities to develop and implement accident-prevention plans. Responsible Care does not require the adoption of any particular accident-prevention technologies, nor does it specify a level of performance that chemical facilities must meet. Instead, this trade association program requires that managers of chemical facilities assess potential accident risks, develop a written emergency response plan, establish a training program for employees, and institute a strategy for communicating risks and plans to the surrounding community (American Chemistry Council 2004).

Both Section 112(r) of the Clean Air Act and the chemical industry's Responsible Care initiative focus directly on changing the internal management of firms. They are both examples of what we call *management-based strategies*. Similar to the planning requirements that the National Environmental Policy Act many years ago imposed on government agencies,[3] management-based strategies are increasingly being used to require or encourage private businesses to adopt specific planning or other management practices aimed at achieving broader social objectives.

Despite growing interest in and use of management-based strategies by policymakers and business leaders in the United States as well as elsewhere in the world, research and policy analysis has tended to lag behind. To be sure, some important work has focused on selected examples of management-based strategies in a variety of policy contexts (Bardach and Kagan 1982; Braithwaite 1982; Rees 1988; Orts 1995; Gunningham 1996; Parker 2002; Coglianese and Lazer 2003). Yet surprisingly little effort has been made to synthesize the results from the relatively few empirical studies that have been done, or even to characterize management-based strategies as a distinct policy approach deserving of its own serious analysis and consideration. The collection of empirical and theoretical analyses contained in this book therefore brings much-needed focus and attention to management-based strategies. It does so not only by chronicling the emergence of management-based strategies and showing the variation such strategies can take, but by explicitly considering what these strategies can—and cannot—achieve in terms of protecting the environment from the negative effects of business activity.

Management-Based Strategies: A New Approach

Traditional environmental regulation has gone far toward addressing a wide range of environmental problems in the United States (Bok 1996; Davies and Mazurek 1998; Portney and Stavins 2000). Since the 1970s, overall levels of air quality have improved steadily even while economic activity and vehicle miles traveled have increased dramatically (U.S. EPA 2003a).[4] The number of polluted rivers, streams, and lakes has declined, and swimming, fishing, and boating are now possible in waters that once functioned essentially as open sewers (U.S. EPA 2003a). Abandoned hazardous waste sites have been cleaned up and contaminated land has been restored to productive use (U.S. EPA 2003a).[5] Due in part to traditional environmental regulation, many large and obvious environmental problems have been greatly reduced, if not eliminated altogether (Coglianese 2001a).

Notwithstanding the overall improvement in various environmental conditions in the United States, many pressing and complex environmental problems remain, and new problems loom. Air pollution from motor vehicles, electric power plants, and other sources continues to raise concern about unhealthy air quality in metropolitan areas, degradation of the protective ozone layer, and global climate change. Many lakes and rivers remain polluted, especially from runoff of excess pesticides from agricultural land and toxic contaminants from other diffuse, non-point sources. Rapid development and expanding urbanization threaten biodiversity and ecosystem resilience. The acknowledged sources of environmental degradation have expanded beyond large industrial operations to include retail establishments, universities, hospitals, farms, and even office buildings where government and other service professionals work.

To respond both to lingering and emerging environmental challenges, management-based strategies appear to have an important role to play in the future of environmental protection (Elliott 1997). Because they focus on what goes on inside organizations, management-based strategies can be adapted to a variety of circumstances and appear well suited to the particular challenges characteristic of many of today's remaining environmental problems.

Many of the most intractable environmental problems have diffuse origins. Often there is no smokestack or discharge pipe where a pollution control device can be mandated or installed (Kettl 2002). Furthermore, some of the most significant remaining sources of pollution today are more heterogeneous than the sources of the past, compounding the information burden on government if it is to rely on conventional regulatory tools. In addition, some of today's most serious environmental problems grow out of complex organizational and technological interactions that can be directly affected by management (Perrow 1984).

Management-based strategies are appealing because they can give firms the flexibility to develop their own solutions and plans, a considerable advantage given the complex, heterogeneous causes of various environmental problems. Unlike technology-based standards that can be directed at the problems of entire industry sectors, management-based strategies are adaptable to a wide variety of individual settings, regardless of the size, age, or operational type of facilities. Management-based strategies can leverage the informational advantage of managers within business organizations, enlisting them to identify ways to solve the specific problems created by their facilities' operations (Coglianese and Lazer 2003).

Consider the environmental problems created by the nation's vast network of university and industry research laboratories. These laboratories generate diverse chemical wastes left over from the many different experiments run by faculty and student researchers, and they pose an array of risks of spills and accidents. Crafting technology- or performance-based standards to address pollution from a vast and diverse network of laboratories would require an enormous amount of fine-grained—and ever-changing—information. It may be far better to have those who operate these facilities develop management systems that address the unique hazards created by their operations (U.S. EPA 2002).

Policymakers and business leaders increasingly recognize that what goes on inside the black box of the organization is of critical importance for overall environmental quality. The size of a regulated entity's environmental footprint is neither necessarily nor completely determined by the raw materials it uses or the products or services it produces. An organization's impact on the environment is also based on the actions of its managers: the information they collect and share, the goals they develop and articulate, the types of issues to which they give attention, the reporting structures they establish, the kind of professionals they include in key meetings, the aspects of performance they monitor, and the way they factor environmental performance into employee rewards and penalties.

Management-Based Strategies and the Policy Toolbox

Given the challenging nature of contemporary environmental problems, it is not surprising that government policymakers are taking increasing interest in management-based strategies. In addition to the examples already mentioned and discussed elsewhere in this book, anyone who reviews the *Federal Register* carefully will find a variety of other ways that environmental regulators are working to affect the management of organizations. For example, to reduce contaminated runoff from farms where large numbers of animals are concentrated, the U.S. Environmental Protection Agency (EPA) requires that farm operators develop nutrient management plans that detail how manure will be stored, transported, and used so as to protect clean water sources (U.S. EPA 2003b). A recent EPA proposal to reduce the foul odors and other impacts associated with sewage sludge would encourage municipalities to implement environmental management systems and use third parties to audit their management practices (U.S. EPA 2003c).

The emergence of these and other similar proposals should lead scholars and policy analysts to consider what role management-based strategies should play in the overall environmental policy toolbox. Yet, at least until now, management-based strategies have received remarkably scant consideration in the environmental policy literature. The question of what regulatory instrument to use is usually posed as a choice between technology- and performance-based standards (Breyer 1982; Viscusi 1983). Technology standards, also sometimes called "design" or "specification" standards (Bohm and Russell 1985; U.S. Congress 1995), specify a particular *means* that firms must use to control pollution. Performance standards, also sometimes called "harm-based standards" (U.S. Congress 1995), specify a particular *end* that firms must achieve in their environmental programs (Project on Alternative Regulatory Approaches 1981; Coglianese et al. 2003). Together, technology and performance standards constitute most of traditional environmental regulation, and they have generated considerable discussion within the legal and policy literature (U.S. EPA 1990; Callan and Thomas 1996; Blackman and Harrington 1998; Driesen 1998; Wagner 2000; Morag-Levine 2003).

Market-based environmental policy instruments also play an important role in the policymaker's toolbox and have received substantial attention in the scholarly literature (Ackerman and Stewart 1985; Hahn and Hester 1989; Stavins 1998, 2003). Market-based instruments establish explicit or implicit prices on environmental pollution (Blackman and Harrington 1998). They attempt to structure incentives so that private sector managers internalize the costs of their environmental impacts in more socially optimal ways. Examples of market-based instruments include emissions fees that require firms to pay when they emit pollution (Dales 1968; Rose-Ackerman 1973; Majone 1976; DOE 1989; Hahn 1989; U.S. EPA 1990; U.S. Congress 1995), input taxes that charge firms for the volume of materials or chemicals they use in their production (Stahr 1971; Baumol and Oates 1979; Hahn 1989; Blackman and Harrington 1998), and marketable permits that allow firms to trade or sell emissions credits with others (Bohm and Russell 1985; Hahn 1989; U.S. EPA 1990; Stavins 1991; Hahn and Stavins 1992; Schmalensee et al. 1998; Ellerman et al. 2000; Stavins 2003).

In addition to traditional and market-based regulation, a few other tools in the environmental policy toolkit have received attention in the legal and policy literature. These include the direct expenditure of government dollars to achieve policy goals (Stahr 1971), such as government spending to construct water treatment plants (Blackman and Harrington 1998) or to preserve or restore endangered ecosystems (Bohm and Russel 1985). Information disclosure programs—such as the Toxics Release Inventory (TRI)—have also received considerable discussion in the literature (Hamilton 1995; Kleindorfer and Orts 1998; Sunstein 1999; Karkkainen 2001; Graham 2002; Hamilton 2005; Fung et al. 2006). Finally, as federal and state governments have launched dozens of voluntary environmental programs to encourage energy-efficient lighting, promote waste reduction, and foster water conservation, among other goals, a substantial literature has developed around the establishment of these voluntary programs (Arora and Cason 1996; Mazurek 1998; Delmas and Terlaak 2001; Khanna 2001; Prakash 2001; Potoski and Prakash 2002; de Bruijn and Norberg-Bohm 2005).

Although all policy instruments or strategies can have an impact on the internal management of firms, any such effects are only indirect. For example, when Congress in 1986 required firms to report publicly their releases of toxic substances, the main purpose of this requirement was to inform employees and the public, and then at best indirectly to catalyze external forces to bring pressure on firms' internal management to reduce toxic emissions (Karkannien 2001). After the release of the first required reports, some private sector managers were shocked to learn that their company's releases put them at or near the top of EPA's list of polluters. In these managers' firms, information disclosure triggered intense management attention, goal-setting, training, information collection, and monitoring (Graham 2000). Yet this was only a side effect of the TRI law, albeit a positive one; it was not the law's main purpose. In fact, none of the instruments discussed in the general environmental policy literature aim *directly* at influencing specific *management* practices.[6]

The closest parallel to management-based strategies can be found in the so-called reflexive environmental policies discussed in the legal literature (Teubner et al. 1994; Orts 1995; Fiorino 1999).[7] Rather than impose direct controls on behavior, reflexive strategies enlist "intermediary institutions" (Orts 1995, *1264*) such as companies, trade associations, and standards organizations "to encourage thinking and behavior in the right direction" (Orts 1995, *1264*; see also Ayres and Braithwaite 1992). Reflexive law recognizes that the complexity of environmental problems limits the ability of regulators to design and implement effective policies. Government therefore should foster conditions that encourage managers of regulated entities to identify and pursue environmental improvement opportunities.

The idea of reflexive law originated in Europe, and its most prominent application is the European Eco-Management and Audit Scheme (EMAS). EMAS is a voluntary standard established by the European Union to recognize facilities that comply with environmental laws, implement environmental management systems, and publish independently verified reports on their environmental performance (European Commission 2005). EMAS encourages managers to adopt a critical approach to their environmental conduct through environmental management planning, review, and disclosure (Orts 1995).

In our view, EMAS is also an excellent example of a management-based strategy, since it aims directly to shape firms' environmental management. We believe such management-based strategies merit their own place in the U.S. policy lexicon as they may prove to be, at least in certain important cases, valuable tools for environmental protection. Yet management-based strategies themselves can take different forms and can be deployed by different kinds of institutions. To analyze their role in the policy toolkit, it is helpful to consider four major types of management-based strategies.

Types of Management-Based Strategies

Both *governmental* and *nongovernmental* institutions can deploy management-based strategies. These strategies can also either *mandate* management practices or simply *encourage* them. Consider the following four examples of management-based strategies in use today.

Massachusetts' Toxic Use Reduction Act

Since 1989, the Commonwealth of Massachusetts has addressed the risks from toxic chemicals with a management-based regulation known as the Toxics Use Reduction Act. The origins of this law can be traced to a cluster of childhood leukemia cases found in Woburn, Massachusetts, that emerged during the early 1980s and that many residents attributed to toxic contamination. The ensuing public concern about environmental degradation, and hazardous wastes in particular, bolstered environmental activists in the state, including the Massachusetts Public Interest Research Group (MassPIRG). By the late 1980s, it appeared likely that MassPIRG would use its clout to gain passage of a law banning or phasing out a broad range of toxic chemicals.

To stave off the imposition of chemical bans, industry entered into negotiations with MassPIRG. These negotiations resulted in the unanimous approval by the legislature of a management-based regulation requiring firms to analyze the use and flow of toxic chemicals throughout their facilities, develop plans to reduce their use of these chemicals, and submit reports of their planning to the state. Adopted in 1989, the Toxics Use Reduction Act (TURA) also requires a state-authorized pollution-prevention planner to certify that each plan has met the criteria in TURA for a rigorous toxics use reduction plan. TURA does not, however, require firms to comply with their own internal plans, nor does it require them to meet any individual performance targets. The law simply asks managers to study their use of toxic chemicals and engage in a planning process designed to identify possible strategies for pollution prevention. Today, 14 states have laws similar to TURA that encourage the prevention of toxic pollution through required management efforts.

Ford Motor Company's ISO 14001 Supplier Requirement

Major corporations seeking to improve the environmental performance of their operations face enormous challenges in overseeing diverse and geograph-ically disparate manufacturing facilities. Before the mid-1990s, Ford Motor Company, like many large manufacturing organizations, functioned as a loose federation of facilities that operated mostly independently. By the late 1990s, however, Ford had become a global organization with all of its national facili-ties operating under a single umbrella. Coordinating the environmental performance of all of these plants represented a major challenge, as the Ford organization included 140 manufacturing sites operating under the regulatory regimes in 26 countries (O'Brien 2001).

Rather than attempt to establish a uniform set of pollution control technolo-gies or environmental performance standards at all plants, Ford's corporate headquarters chose a new approach that had been gaining acceptance, partic-ularly in Europe and Asia. Ford decided to impose a uniform set of environmental management practices at all of its plants. By focusing on man-agement, Ford was able to establish a common, disciplined approach to environmental protection while respecting local differences that might call for different control technologies or performance standards.

In 1999, Ford's corporate offices decided to take its mandate for disciplined environmental management one step further. Beginning in 2003, its "first-tier" suppliers of tires, seats, pedals, wiring, spark plugs, switches, mufflers, and many other parts used in Ford vehicles would need to adopt the same environ-mental management practices Ford required of its own plants. Ford did not require these suppliers to use particular environmental control technologies or achieve a particular level of environmental performance. Instead, it required them to adopt management practices that met the criteria set forth in ISO 14001, an international environmental management system standard (Ford Motor Company 1999; Wilson 2001).

Other major companies have taken similar steps to improve the environmental management of their facilities and suppliers. For example, on the same day that Ford announced it was imposing environmental management requirements on its suppliers, General Motors Corporation issued a similar requirement for standardized environmental management for its first-tier suppliers (General Motors Corporation 1999).[8]

Federal and State Performance Track Programs

As with major corporations, federal, state, and local governments in the United States also face a challenge in trying to see that the thousands of industrial facilities in their jurisdictions reduce their impact on the environment. For decades, the strategies governments used were regulatory, but in recent years many state and federal agencies have initiated programs offering recognition and other incentives in an attempt to encourage firms to achieve voluntary environmental improvements. Perhaps the best example of such a program is the National Environmental Performance Track, initiated by EPA in 2000. Facilities that participate in Performance Track have their names posted on EPA's Web site, are invited to meetings with high-ranking EPA officials, and receive exemptions from routine agency inspections and certain regulatory requirements.

To qualify for entry into Performance Track, a facility must comply with all environmental laws and demonstrate a record of environmental accomplishments that go beyond what the law requires. The facility must show that it has an environmental management system in place and that this system has been verified by an independent third party. It must also show that it is open to engagement with the community and that it has made a commitment to improve environmental performance in four areas.[9] Although EPA provides guidance to managers about the types of commitments it expects—for example, water or energy conservation—managers may improve at their own pace, in their own ways. The important factor for EPA is that managers commit somehow to reducing their ecological footprint and that they develop a management system for achieving those reductions.

State regulatory agencies, including those in Michigan, Texas, Virginia, and Wisconsin, have adopted programs similar to Performance Track that offer recognition and other incentives to facilities with strong environmental management programs. All of these programs encourage, but do not require, firms to adopt management systems that may help them reduce environmental impacts further than the levels mandated by conventional environmental regulation.

Portland Cement Association's Sustainability Program

Like large firms and government environmental agencies, trade associations also have an interest in promoting strong environmental performance. A well-publicized environmental problem at one firm can drag down the reputation of an entire industry and prompt government to adopt further environmental regulation of the industry. Many trade associations have attempted to promote strong environmental performance by establishing codes of practices for their members. Rather than establish performance- or technology-based standards, these codes often tend to focus on members' management activities.

Portland Cement Association's Sustainability Program is a case in point. In 1991, the trade association's board of directors adopted seven principles that call on member companies to promote overall safety and health, manage wastes responsibly, and improve energy efficiency. The trade association did not advocate any specific technologies or performance standards for its members, but rather encouraged members to develop their own performance measures and chart progress toward their achievement (PCA 2004a).

In 2004, the trade association's executive committee went a step further, encouraging members to adopt formal environmental management systems. PCA promoted the benefits of environmental management systems (EMSs) in its publications and offered EMS training. It established a goal of having verifiable EMSs in place at 90 percent of U.S. cement plants by the end of 2020. It pledged to recognize firms that implemented environmental management systems through an awards program. In addition, the trade association now collects information about members' environmental performance and plans to develop benchmarks for energy consumption and other performance measures so that managers can compare their performance with the industry norm (PCA 2004b).

A Typology of Management-Based Strategies

As these examples show, management-based strategies can take a variety of forms and can be adopted by a variety of organizations, including government agencies, private firms, and trade associations. Although each is different, these four examples share a clear focus on management itself. The strategies all directly seek to influence the attention, information, authority, and financial resources of managers toward the achievement of environmental improvements. The strategies do not necessarily require managers to achieve any specific outcomes, but they do allow them the flexibility to choose their own measures to reduce their environmental impacts.

Each of the examples illustrates a different type of management-based strategy. The Massachusetts TURA and similar state laws exemplify management-based regulation. These laws are nondiscretionary imperatives that firms implement specified management practices. Another example of a government-imposed requirement for management-based practices is the risk management planning requirement detailed in Section 112(r) of the federal

Clean Air Act. Firms operating in particular chemical manufacturing and distribution sectors, and which have more than a specified number of employees, are required to undertake the planning and other management practices called for in Section 112(r).

Just as government has required management-based activities under TURA and the federal risk management rule, Ford Motor Company has mandated a set of management practices. Ford requires its suppliers to develop an environmental policy, assess the environmental impacts of their operations, set goals, assign responsibility, train workers, and document progress, as described in ISO 14001. The penalty for failing to adopt an ISO 14001 system is the potential of being dropped from Ford's list of preferred suppliers.[10] Ford's actions to require management activities are not unique. The American Chemistry Council (ACC) (formerly the Chemical Manufacturers Association) has similarly required environmental management practices of its member firms under its Responsible Care program. The ACC now requires that chemical firms implement a "Responsible Care Management System" as a condition for membership in the trade association (ACC 2005). In both the Ford and ACC cases, the management-based regulator is a private organization, not the government. We therefore refer to such private sector requirements as management-based *mandates*, to distinguish them from government-imposed management-based *regulation*.

Mandatory management-based strategies stand in contrast to efforts that simply encourage improved environmental management. EPA's National Environmental Performance Track and similar state programs, such as the Clean Texas Program run by the Texas Natural Resource Conservation Commission, offer incentives to facilities to implement environmental management practices. With these programs, facilities can base their decision of whether to participate on their assessment of the costs and benefits of implementing the management practices government seeks to encourage. We refer to governmental efforts to encourage, but not require, improved environmental management as management-based *incentives*.

The Portland Cement Association's EMS program is also voluntary. Unlike Responsible Care in the chemicals sector, adoption is not a requirement for membership in the cement trade association, even though it is strongly encouraged. The International Organization for Standardization (ISO) is another private organization that exerts some modest pressure on firms to improve environmental management. ISO convenes the process whereby representatives from various countries, private sector firms, and nongovernmental organizations develop and revise environmental management system standards. ISO does not require any firms to comply with its standards, but by bringing parties together to draft management standards it enables and encourages their use (ISO 2005). We refer to such nongovernmental efforts to encourage improvements in environmental management as management-based *pressures*.

	Governmental User	Nongovernmental User
Management Required	*Management-based regulations* Examples: • Risk management planning required under Clean Air Act Section 112(r) • Massachusetts Toxics Use Reduction Act	*Management-based mandates* Examples: • American Chemistry Council's Responsible Care Program • Ford Motor Company's requirement that suppliers become certified to ISO 14001
Management Encouraged	*Management-based incentives* Examples: • U.S. EPA's National Environmental Performance Track • Texas Natural Resource Conservation Commission's Clean Texas Program	*Management-based pressures* Examples: • Portland Cement Association's Cement Manufacturing Sustainability Program • International Organization for Standardization's (ISO) 14001 Standard

FIGURE 1-1. Types of Management-Based Strategies

Figure 1-1 summarizes the major differences in management-based strategies. It distinguishes four strategies—regulations, mandates, incentives, and pressures—based on the type of institution that deploys the strategy and whether the specified management practices are mandated or encouraged.

Management-Based Strategies and Environmental Management Systems

Management-based strategies are related to another emerging environmental practice: the adoption of environmental management systems (EMSs). During the past two decades, thousands of organizations in the United States, Europe, and Asia have adopted EMSs, and the number keeps growing (Coglianese and Nash 2001; Delmas 2002). What we mean by management-based strategies, however, is both broader than and, in important ways, different from EMSs.

EMSs are the internal rules and organizational structures managers use to routinize behavior in order to satisfy their organization's environmental goals. Although the specific features of these systems vary across organizations, under almost any EMS managers establish an environmental policy or plan; implement the resulting plan by assigning responsibility, providing resources, and training workers; check progress through systematic auditing; and report results to top managers.[11] EMSs are effectively "regulatory" structures that arise from within organizations (Coglianese and Nash 2001, *1*).

Management-based *strategies*, in contrast, are used by those who are *outside* an organization to change the management practices and behaviors of those on the *inside*. A major distinction between management-based strategies and EMSs therefore lies in who is requiring or encouraging the improved environmental management. When managers within individual companies devise EMSs for use in their own facilities, they are changing their own behavior or seeking to change the behavior of those individuals whom they oversee within their organization. When government agencies, major customers, or trade associations encourage or require companies to develop EMSs, they are outsiders deploying a management-based strategy.

In an earlier volume, *Regulating from the Inside*, we called attention to the implications of the growing use of environmental management systems for public policy (Coglianese and Nash 2001). Describing the potential for a "management-based environmental policy," we examined an array of policy options for promoting the adoption of EMSs. Specifically, we analyzed the potential for government or private sector firms or trade associations to mandate EMS adoption. We also noted that governments might increase the benefits of EMS adoption through public recognition, enforcement forbearance, and regulatory flexibility; or they might lower the costs of EMS adoption through education, technical assistance, subsidies or tax credits, and audit protection (Coglianese 2001b). All are examples of different management-based strategies.

In the years since the publication of *Regulating from the Inside*, federal and state agencies, trade associations, and private firms have, to varying degrees, pursued each of the policies we identified (Coglianese and Nash 2002). In April 2002, the White House Council on Environmental Quality (CEQ) and Office of Management and Budget (OMB) sent a memorandum to all federal agency heads urging them to comply with Executive Order 13148, which mandates that federal agencies implement EMSs by the end of 2005 (CEQ 2002). In April 2003, the ACC revamped its Responsible Care program to require all trade association members to adopt EMSs that include performance metrics and that are certified by external third parties (Yosie 2003). That year the Coalition for Environmentally Responsible Economies (CERES) launched its Facility Reporting Program to gather information about facility-level environmental management (CERES 2005). In April 2004, EPA issued a new Strategy for Determining the Role of Environmental Management Systems in Regulatory Programs that calls for careful experimentation with using EMSs in regulatory programs (U.S. EPA 2004). EPA has also funded projects in nine states, at a cost of more than $1.5 million, to explore the role of EMSs in permitting and related issues (U.S. EPA 2005c). Finally, EPA's Sector Strategies Program, working in partnership with trade associations, has developed EMS implementation guides for five sectors: agri-business, metal casting, metal finishing, shipbuilding and repair, and specialty-batch chemicals (U.S. EPA 2005e).

What is the likely impact of these strategies? While management-based strategies differ from EMSs with respect to who initiates the effort to change environmentally detrimental behavior, they do share a common focus on management activities. They are both premised on the belief that improvement in environmental management will lead to improvements in environmental outcomes. Just as many managers have implemented EMSs specifically to improve their firm's environmental performance (Nash et al. 1999; Andrews et al. 2001; Florida and Davison 2001), governments, trade associations, and other organizations that deploy management-based strategies do so in order to generate improvement in environment conditions.

To understand the potential of management-based strategies, then, it helps to consider what we know about how EMSs affect the environmental performance of facilities that use them. If the adoption of EMSs leads to an improvement in firms' environmental performance, then perhaps management-based strategies could be an effective means of improving environmental performance too.

A growing body of case study evidence suggests that in many instances EMSs have had positive effects on environmental outcomes. EPA's EMS Web site includes about a dozen case studies of firms that have improved their environmental performance by implementing an EMS. The achievements firms have realized include, among other things, increased recycling, improved water conservation, reduced releases of toxic chemicals, and reduced consumption of material inputs (U.S. EPA 2005b). For example, an Alcoa subsidiary in South Carolina cut waste generation in half after implementing an EMS (Rondinelli and Vastag 2000). EMS adoption helped a Louisiana Pacific wood products plant find a way to recycle wood chips, reducing waste and saving money (Coglianese and Nash 2001).

Although case studies can show that EMSs might improve environmental performance in some firms, they tell us little about how EMSs will work in general. Generalizing from individual case studies is difficult because these studies do not distinguish between the role of the EMS and the role of other factors that might also contribute to improvements in environmental performance. Louisiana Pacific, for example, introduced its EMS after EPA filed suit against the company for unlawful air releases (Coglianese and Nash 2001). Was the EMS the causal factor leading the company to improve environment outcomes? Or was the government's criminal enforcement action the key cause of both the decision to adopt an EMS and to improve environmental outcomes? In the wake of such an enforcement action, Louisiana Pacific's overall heightened commitment to making environmental improvements, rather than its EMS, may have been the critical factor in instituting new, environmentally beneficial practices.

In an effort to overcome the limitations of individual case studies, some researchers have conducted large-scale studies of the impacts of EMSs. Some of the studies suggest that EMS adoption has little discernable impact on environmental performance. Automotive assembly plants that implemented EMSs reportedly have had no better environmental performance in terms of toxic releases, criteria air pollutant emissions, RCRA hazardous wastes, and instances of noncompliance with regulations than those that did not (Matthews 2001; Matthews et al. 2004). British facilities that adopted environmental management systems had no fewer accidents or regulatory enforcement actions than similar facilities without EMSs, although they achieved higher levels of procedural performance such as record keeping and worker training (Dahlstrom et al. 2003).

However, increasingly, the weight of the evidence suggests that EMSs are positively correlated with improvements in environmental performance. Managers of manufacturing plants in Pennsylvania that had implemented EMSs reported greater reductions in air pollution, electricity use, and solid waste than their non-EMS peers (Florida and Davison 2001). Managers of Mexican manufacturing firms reported that their environmental performance improved as they completed more steps toward ISO 14001 adoption (Dasgupta et al. 2000). Electronics facilities with EMSs had somewhat lower releases of toxic emissions than comparable plants that had not adopted formal environmental management systems (Russo 2001). In a sample of Fortune 500 firms, EMS adoption correlated with lower toxic emissions per unit of output, particularly in firms with large toxic releases relative to their size (Anton et al. 2004). Among manufacturing facilities in the United States, the adoption of elements of an EMS is associated with relatively greater reductions in releases of toxic chemicals (King et al. 2005). U.S. facilities that have developed EMSs certified to meet the ISO 14001 standard have tended to reduce harmful air emissions to a greater extent than those that have not (Potoski and Prakash 2005a) and they comply more fully with the Clean Air Act (Potoski and Prakash 2005b). European firms with certified EMSs reportedly undertake more environmental initiatives than those without such systems (Johnstone 2001).

Key Questions about Management-Based Strategies

The accumulated research showing an association between environmental management systems and improved environmental performance strengthens the rationale for analyzing management-based strategies. If EMSs matter, then strategies that aim directly at improving firms' environmental management deserve more systematic attention than they have received to date. How should such an inquiry into management-based strategies proceed? We believe that to understand the appropriate role for management-based strategies and the conditions under which they are appropriate calls for careful assessment of at least five main questions.

The Importance of Management

The first question focuses on the importance of a firm's management to its environmental performance. As we have just noted, many studies of EMS adoption suggest that firms can improve their environmental performance by making management improvements. Yet improvements are by no means certain for all firms under all conditions (Andrews et al. 2001; Coglianese and Nash 2001; Nash and Ehrenfeld 2001; Speir 2001). Since management-based strategies seek to influence management, they will be appropriate only if and when management itself is an important factor causally related to resource consumption, waste production, and other factors that degrade the environment. To make the most of management-based strategies, the relationship between a firm's environmental performance and its management (independent of regulatory, economic, and social factors) still needs to be better understood (Gunningham et al. 2003).

The Impact of Management-Based Strategies

Assuming that management is a key factor affecting certain environmental outcomes, the second question is whether management-based strategies actually change what managers do. Do management-based strategies—that is, actions taken by outsiders to influence firms' internal management—successfully permeate the black box of the firm and influence organizational structure and decisionmaking in a positive way? Or do managers mostly ignore these strategies or game outsiders with symbolic gestures that lack substantive results? For example, when Ford Motor Company mandated that its suppliers implement ISO 14001, did this requirement actually lead to significant changes in suppliers' practices? Or did suppliers simply go through the motions, establishing management plans that looked good on paper but did not result in genuine change?

Strategic Design

A third question to ask about management-based strategies is whether some strategies are more effective than others. Presumably some do work better than others, at least in certain circumstances. If so, then we need to identify which types of strategies yield the most successful outcomes under which conditions, and we need to understand why. Management-based strategies can vary in the incentives they offer, the amount of direction they provide, and the oversight mechanisms they deploy. What types of incentives work to motivate improvements in management practices?

Government agencies have already tried a variety of incentives, including public recognition, inspection forbearance, and regulatory flexibility. Participants in these programs have advocated for still greater incentives, such as expedited permitting, streamlined reporting, and green investment ratings to motivate facilities to participate (ECOS 2005). Are these additional incentives necessary?

How much discretion should outsiders give to insiders? For example, some management-based regulations call for firms to do little more than adopt "appropriate" management plans, while others require plans that meet detailed and extensive criteria. Which approach works better?

Similarly, different management-based efforts encourage or require different oversight mechanisms. Many management-based strategies include a requirement that firms engage with external constituencies (ACC 2004; MSWG 2004; U.S. EPA 2005a). Others require the firm to have its management system certified by a third party. Still others require the collection and public disclosure of information (Metzenbaum 2001). How, if at all, do these differences in oversight mechanisms shape firms' environmental management and their overall performance? By observing the impacts of management-based strategies with different methods of oversight, along with other variations in design, we will be able to learn how to deploy management-based strategies more effectively.

Management-Based Strategies and Conventional Regulation

A fourth question centers on the relationships between management-based strategies and existing environmental regulation. Some research suggests that management-based strategies can help firms come into greater compliance with existing regulations through better planning and auditing (U.S. EPA 2005d). Are management-based strategies merely complements to other strategies, strengthening conventional regulation but not providing a substitute for it? Or do management-based strategies offer a third way of achieving environmental protection that is truly independent? As Andrews, Hutson, and Edwards point out in Chapter 5, management-based strategies may lead companies to reduce environmental impacts not currently addressed by government regulation. On the other hand, the existence of extensive government regulation may well constrain the full impact of management-based strategies, because firms would be compelled to follow the law even if the government-mandated action is less effective than alternative ways of managing and reducing their environmental impacts.

Evaluating Management-Based Strategies

The fifth question focuses on how to define and measure the effectiveness of management-based strategies. "Management" by itself is probably too broad a concept for evaluative purposes. Progress in assessing the correlation between management and environmental performance requires researchers to define management in concrete terms using observable characteristics. Without such measures, researchers may continue to conflate managers' commitment or values with their management actions, even though these are likely to be separate factors affecting environmental performance (Coglianese and Nash 2001). If managers' commitment is what matters most in affecting firms' environmental performance, then requiring or encouraging them to implement specific management actions may well not lead to expected performance improvements. Researchers evaluating management-based strategies should be mindful of all the different factors affecting firms' management and develop measures to control for them as much as feasible.

Road Map of the Book

These five questions about management-based strategies form the foundation for this book as much as they map out an agenda for future research. They are also five crucial questions for public policy because, although management-based approaches seem intuitively appealing, we know relatively little about the conditions under which they work, which designs are most effective, and how these approaches fit within a larger environmental policy toolbox. The only way to answer these questions is by conducting empirical research. Toward that end, we have commissioned leading scholars from a variety of disciplinary backgrounds to examine these questions in the light of prominent examples of management-based strategies being used in the public and private sector today. The research these scholars produced, and which is collected in this book, promises the most comprehensive effort to date to define, explain, and evaluate the role of management-based strategies in environmental protection.

Part I of this book focuses on the question of whether management is the right place to look for improvements in environmental performance. Chapter 2, written by political scientist Robert A. Kagan, explores the complex role that management plays in explaining variation in firms' environmental performance. Based on findings from a study of pulp and paper mills in the United States, Canada, Australia, and New Zealand, Kagan suggests that "management style" does influence environmental performance, particularly in shaping how organizational decisionmakers interpret regulatory, market, and community signals. His answer to the initial question of whether management matters is therefore decidedly "yes," even if internal management does not completely determine a firm's actions with respect to the environment.

In Part II, we consider the role of public and private sector mandates to adopt management-based approaches. In Chapter 3, economist Lori Snyder

Bennear reports results from an empirical evaluation of state laws requiring managers to develop pollution prevention plans. In Chapter 4, economist Paul R. Kleindorfer presents findings from a study of EPA's requirement that chemical plants develop risk management plans. In Chapter 5, political scientist Richard N. L. Andrews and his coauthors share findings from a study of management-based mandates imposed by industrial customers on their suppliers. These researchers suggest that requiring specific management practices can indeed motivate performance, but their work also cautions that such mandates by no means guarantee improvements across the board. And in Chapter 6, Howard Kunreuther, Shelley H. Metzenbaum, and Peter Schmeidler provide an analysis of a different and novel kind of mandate—mandatory insurance—as a potential driver for motivating improved environmental management. They postulate that this approach may have great success in motivating effective environmental management.

In Part III, we examine the effectiveness of nonmandatory management-based incentives and pressures. In Chapter 7, Jason Scott Johnston, a lawyer and economist, reports results from an empirical study of the effectiveness of EPA's Strategic Goals Program, a management-based voluntary program established between EPA and the metal-finishing industry in 1998. In Chapter 8, economists Tapas K. Ray and Kathleen Segerson document the impact of an EPA management-based effort called the Clean Charles Initiative, through which government served as the standard-bearer of an ecosystemwide performance management initiative. In Chapter 9, Andrew A. King, an expert on management and organizational behavior, explores the value of management-based strategies to environmental organizations seeking to promote innovative industrial practices. Together, the research in Part III suggests a less sanguine view of management-based strategies, at least those that lack sufficient incentives behind them. The cases examined here are by no means exhaustive, but they do appear to support the view that such strategies are insufficient substitutes for more compulsory efforts to achieve environmental protection.

The research in this book begins to answer the critical questions about the impact of management-based approaches, the effectiveness of different designs, the relationship between management-based strategies and traditional environmental regulation, and the best ways to evaluate these approaches. Some management-based strategies can yield positive results for environmental protection, especially in inducing firms to make improvements in areas that are difficult for government to regulate or that involve complex interactions between people and industrial processes. Yet the results from management-based strategies will not always be dramatic, particularly when programs are designed to encourage rather than require improved environmental management. Management-based strategies are certainly not appropriate for all problems. In many cases, they will serve best to complement, rather than substitute for, other environmental protection strategies.

While we should have no illusions that management-based strategies will become the mainstay of society's approach to environmental protection, this book shows quite well that they can offer measurable and positive societal benefits. Management-based strategies have made a difference in some cases and represent an important approach that government, the private sector, and non-governmental organizations increasingly use, especially when attempting to address some of the most intractable kinds of environmental problems. Those trying to protect the environment would therefore do well to consider more carefully strategies that seek to leverage private sector management to benefit the overall public welfare.

Notes

1. For background on the development of the risk management planning rule under the 1990 Clean Air Act Amendments, see Makris (1998).

2. Subsequently, the Chemical Manufacturers Association changed its name to the American Chemistry Council.

3. The National Environmental Policy Act of 1969, 42 U.S. Code Section 4321–4347. See Caldwell (1998), Karkkainen (2002), and Connaughton (2003).

4. However, this is not to say that all of the reported declines in ambient air pollution can properly be attributed to environmental regulation. For a recent program evaluation of the Clean Air Act analyzing the impact of ambient air quality standards, see Greenstone (2004).

5. This is not to say that the nation's hazardous waste laws are without any problems. For in-depth assessments of Superfund, see Revesz and Stewart (1995), Hamilton and Viscusi (1999), and Probst et al. (2001).

6. In a comprehensive article on environmental policy tools, for example, Kenneth Richards (2000) reviewed 15 major studies on instrument choice, none of which included anything like the management-based strategies we address in this book.

7. In addition, several scholars have also called attention to enforced or mandated self-regulation, which also are closely related to management-based regulation (Bardach and Kagan 1982; Ayres and Braithwaite 1992; Hutter 2001).

8. In Chapter 5, Richard N. L. Andrews and co-authors discuss the environmental performance of automotive suppliers subject to automakers' ISO 14001 requirements.

9. Under existing program guidelines, small facilities may commit to improve in two areas instead of four.

10. See Hutson (2004) for a discussion of Ford's enforcement of this requirement.

11. Committees convened under the auspices of the International Organization for Standardization have drafted ISO 14001, an international standard for environmental management systems. Many firms have used this standard as the template for the design and operation of their own EMSs.

References

ACC (American Chemistry Council). 2004. *Community Awareness and Emergency Response Code of Management Practices*. RCMS Implementation Guidance Appendices, Appendix F: Responsible Care Codes of Management Practices. Washington, DC: American Chemistry Council.

————. 2005. *Responsible Care: The Chemical Industry's Performance*. http://www.-responsiblecare-us.com/about.asp (accessed October 11, 2005).

Ackerman, Bruce, and Richard Stewart. 1985. Reforming Environmental Law. *Stanford Law Review* 37: 1333–65.

Andrews, Richard N. L., Nicole Darnall, Deborah Rigling Gallagher, Suellen Terrill Keiner, Eric Feldman, Matthew L. Mitchell, Deborah Amaral, and Jessica D. Jacoby. 2001. Environmental Management Systems: History, Theory, and Implementation Research. In *Regulating from the Inside: Can Environmental Management Systems Achieve Policy Goals?*, edited by Cary Coglianese and Jennifer Nash. Washington, DC: Resources for the Future Press, 31–60.

Anton, Wilma Rose Q., George Deltas, and Madhu Khanna. 2004. Incentives for Environmental Self-Regulation and Implications for Toxic Releases. *Journal of Environmental Economics and Management* 48 (1): 632–54.

Arora, Seema, and Timothy N. Cason. 1996. Why do Firms Volunteer to Exceed Environmental Regulation? Understanding Participation in EPA's 33/50 Program. *Land Economics* 72 (4): 413–32.

Ayres, Ian, and John Braithwaite. 1992. *Responsive Regulation: Transcending the Deregulation Debate*. New York: Oxford UP.

Bardach, Eugene, and Robert Kagan. 1982. *Going by the Book: The Problem of Regulatory Unreasonableness*. Philadelphia: Temple UP.

Baumol, William J., and Wallace E. Oates. 1979. *Economics, Environmental Policy, and the Quality of Life*. Englewood Cliffs, NJ: Prentice-Hall.

Blackman, Allen, and Winston Harrington. 1998. Using Alternative Regulatory Instruments to Control Fixed Point Air Pollution in Developing Countries: Lessons from International Experience. Discussion paper 98-21. Washington, DC: Resources for the Future.

Bohm, Peter, and Clifford S. Russell. 1985. Comparative Analysis of Alternative Policy Instruments. In *Handbook of Natural Resource and Energy Economics*, edited by Allen V. Kneese and James L. Sweeney. New York: North Holland, 395–460.

Bok, Derek Curtis. 1996. *The State of the Nation: Government and the Quest for a Better Society*. Cambridge, MA: Harvard UP.

Braithwaite, John. 1982. Enforced Self-Regulation: A New Strategy for Corporate Crime Control. *Michigan Law Review* 80: 1466–507.

Breyer, Stephen G. 1982. *Regulation and its Reform*. Cambridge, MA: Harvard UP.

Caldwell, Lynton Keith. 1998. *The National Environmental Policy Act: An Agenda for the Future*. Bloomington: Indiana UP.

Callan, Scott J., and Janet M. Thomas. 1996. *Environmental Economics and Management: Theory, Policy, and Applications*. Chicago: Irwin.

CEQ (Council on Environmental Quality). 2002. Memoranda. http://www.white-house.gov/ceq/ems.html (accessed August 9, 2005).

CERES (Coalition for Environmentally Responsible Economies). 2005. Facility Reporting Project. http://www.ceres.org/sustreporting/frp.php (accessed November 16, 2005).

Coglianese, Cary. 2001a. Social Movements, Law, and Society: The Institutionalization of the Environmental Movement. *University of Pennsylvania Law Review* 150: 85–118.

———. 2001b. Policies to Promote Systematic Environmental Management. In *Regulating from the Inside: Can Environmental Management Systems Achieve Policy Goals?* edited by Cary Coglianese and Jennifer Nash. Washington, DC: Resources for the Future Press, 181–97.

Coglianese, Cary, and David Lazer. 2003. Management-Based Regulation: Prescribing Private Management to Achieve Public Goals. *Law and Society Review* 37 (4): 691–730.

Coglianese, Cary, and Jennifer Nash. 2001. *Regulating from the Inside: Can Environmental Management Systems Achieve Policy Goals?*, Washington, DC: Resources for the Future Press.

———. 2002. Policy Options for Improving Environmental Management. *Environment* 44(9): 13–22.

Coglianese, Cary, Jennifer Nash, and Todd Olmstead. 2003. Performance-Based Regulation: Prospects and Limitations in Health, Safety, and Environmental Protection. *Administrative Law Review* 55(4): 705–30.

Congressional Record. 1989. 101st Cong., 1st sess., vol. 135, pt. 75. Introduction of the Air Toxics Control Act of 1989.

Connaughton, James L. 2003. Modernizing the National Environmental Policy Act: Back to the Future. *New York University Environmental Law Journal* 12(1): 1–17.

Dahlstrom, Kristina, Chris Howes, Paul Leinster, and Jim Skea. 2003. Environmental Management Systems and Company Performance: Assessing the Case for Extending Risk-Based Regulation. *European Environment* 13: 187–203.

Dales, John H. 1968. *Pollution, Property and Prices.* Toronto: University of Toronto Press.

Dasgupta, Susmita, Hemamala Hettige, and David Wheeler. 2000. What Improves Environmental Performance? Evidence from Mexican Industry. *Journal of Environmental Economics and Management* 39(1): 39–66.

Davies, J. Clarence, and Jan Mazurek. 1998. *Pollution Control in the United States: Evaluating the System.* Washington, DC: Resources for the Future Press.

de Bruijn, Theo, and Vicki Norberg-Bohm (eds.). 2005. *Industrial Transformation: Environmental Policy Innovation in the United States and Europe.* Cambridge, MA: MIT Press.

Delmas, Magali A. 2002. The Diffusion of Environmental Management Standards in Europe and in the United States: An Institutional Perspective. *Policy Sciences* 35(1): 1–119.

Delmas, Magali A., and Ann K. Terlaak. 2001. A Framework for Analyzing Environmental Voluntary Agreements. *California Management Review* 43 (3): 44–66.

DOE (Department of Energy). 1989. A Compendium of Options for Government Policy to Encourage Private Sector Responses to Potential Climate Changes: Methodological Justification and Generic Policy Instruments. Rep. No. DOE/EH-

0103. Washington, DC: U.S. Department of Energy.

Driesen, David M. 1998. Is Emissions Trading an Economic Incentive Program?: Replacing the Command and Control/Economic Incentive Dichotomy. *Washington & Lee Law Review* 55(2): 289–350.

ECOS (Environmental Council of the States). 2005. Survey of State Support for Performance-Based Environmental Programs and Recommendations for Improved Effectiveness. http://www.epa.gov/performancetrack/downloads/ECOS_Report_Final_01-13-05.pdf (accessed June 28, 2005).

Ellerman, A. Denny, Paul L. Joskow, Richard Schmalensee, Juan-Pablo Monterro, and Elizabeth M. Bailey. 2000. *Markets for Clean Air: The U.S. Acid Rain Program.* Cambridge, UK: Cambridge UP.

Elliott, E. Donald. 1997. Toward Ecological Law and Policy. In *Thinking Ecologically: The Next Generation of Environmental Policy,* edited by Marian R. Chertow and Daniel C. Esty. New Haven: Yale UP, 170–88.

European Commission. 2005. EMAS – The Eco-Management and Audit Scheme. http://www.europa.eu.int/comm/environment/emas/about/summary_en.htm (accessed August 4, 2005).

Fiorino, Daniel J. 1999. Rethinking Environmental Regulation: Perspectives on Law and Governance. *Harvard Environmental Law Review* 23: 441–69.

Florida, Richard, and Derek Davison. 2001. Why Do Firms Adopt Advanced Environmental Practices (And Do They Make a Difference?). In *Regulating from the Inside: Can Environmental Management Systems Achieve Policy Goals?,* edited by Cary Coglianese and Jennifer Nash. Washington, DC: Resources for the Future Press, 82–104.

Ford Motor Company. 1999. Ford Becomes First U.S. Automaker to Require Suppliers to Achieve ISO 14001 Certification. Sept. 21, 1999. Press release.

Fung, Archon, David Weil, Mary Graham, and Elena Fagota. 2006. The Effectiveness of Regulatory Disclosure Policies. *Journal of Policy Analysis and Management* 25: 155.

General Motors Corporation. 1999. General Motors Sets New Level of Environmental Performance for Suppliers (Sept. 21). Press release.

Graham, Mary. 2000. Regulation by Shaming. *Atlantic Monthly* 285(4): 36–40.

———. 2002. *Democracy by Disclosure: The Rise of Technopopulism.* Washington, DC: Brookings Institution Press.

Greenstone, Michael. 2004. Did the Clean Air Act Cause the Remarkable Decline in Sulfur Dioxide Concentrations? *Journal of Environmental Economics and Management.* 47: 585–611.

Gunningham, Neil. 1996. From Compliance to Best Practice in OHS: The Role of Specification, Performance, and Systems-Based Standards. *Australian Journal of Labor Law* 9: 221–46

Gunningham, Neil, Robert A. Kagan, and Dorothy Thornton. 2003. *Shades of Green: Business, Regulation, and the Environment.* Stanford, CA: Stanford University Press.

Hahn, Robert W. 1989. *A Primer on Environmental Policy Design.* New York: Harwood Academic Publishers.

Hahn, Robert W., and Gordon L. Hester. 1989. Marketable Permits: Lessons for Theory and Practice. *Ecology Law Quarterly* 16: 361–406.

Hahn, Robert W., and Robert N. Stavins. 1992. Economic Incentives for

Environmental Protection: Integrating Theory and Practice. *The American Economic Review.* 82(2): 464–68.

Hamilton, James T. 1995. Pollution as News: Media and Stock Market Reactions to the Toxics Release Inventory Data. *Journal of Environmental Economics and Management* 28(1): 98–113.

———. 2005. *Regulation through Revelation: The Origin, Politics, and Impacts of the Toxics Release Inventory Program.* Cambridge, UK: Cambridge UP.

Hamilton, James T., and W. Kip Viscusi. 1999. *Calculating Risks: The Spatial and Political Dimensions of Hazardous Waste Policy.* Cambridge, MA: MIT Press.

Hutson, Andrew. 2004. Diffusion of Environmental Practices through Value Chain Relationships in the United States and Mexico. Paper presented at the 26th Annual APPAM Research Conference. October 28–30, 2004, Atlanta, GA.

Hutter, Bridget. 2001. *Regulation and Risk: Occupational Health and Safety on the Railways.* Oxford: Oxford UP.

ISO (International Organization for Standardization). 2005. Overview of the ISO System. http://www.iso.org/iso/en/aboutiso/introduction/index.html (accessed August 9, 2005).

Johnstone, Nick. 2001. The Firm, the Environment, and Public Policy. Working Party on National Environmental Policy, Organisation for Economic Co-operation and Development. ENV/EPOC/WPNEP(2001)31/Final. Paris: OECD.

Karkkainen, Bradley C. 2001. Information as Environmental Regulation: TRI and Performance Benchmarking, Precursor to a New Paradigm? *Georgetown Law Journal* 89: 257–370.

———. 2002. Toward a Smarter NEPA: Monitoring and Managing Government's Environmental Performance. *Columbia Law Review* 102 (4): 903–74.

Kettl, Donald F. 2002. *Environmental Governance: A Report on the Next Generation of Environmental Policy.* Washington, DC: Brookings Institution Press.

Khanna, Madhu. 2001. Non-Mandatory Approaches to Environmental Protection. *Journal of Economic Surveys* 15(3): 291–324.

King, Andrew A., Michael J. Lenox, and Ann Terlaak. 2005. The Strategic Use of Decentralized Institutions: Exploring Certification with the ISO 14001 Management Standard. *Academy of Management Journal* 48(6): 1091–106.

Kleindorfer, Paul R., and Eric Orts. 1998. Informational Regulation of Environmental Risks. *Risk Analysis* 18(2): 155–70.

Lodge, George C., and Jeffrey F. Rayport. 1991. Responsible Care. Harvard Business School Case #9-391-135. Boston: Harvard Business School.

Majone, Giandomenico. 1976. Choice among Policy Instruments for Pollution Control. *Policy Analysis* 2: 589–613.

Makris, Jim. 1998. EPA Perspective on Advances in Process Safety. Transcript of Keynote Address presented at the 1st Annual Symposium of the Mary Kay O'Connor Process Safety Center, "Beyond Regulatory Compliance, Making Safety Second Nature." Mary Kay O'Connor Process Safety Center, Texas A & M University, College Station, Texas. March 1998. http://psc.tamu.edu/symposium/1998/Papers/Makris.htm (accessed April 1, 2005).

Matthews, Deanna Hart. 2001. Assessment and Design of Industrial Environmental

Management Systems. PhD dissertation. Pittsburgh, PA: Carnegie-Mellon University.

Matthews, Deanna Hart, Chris T. Hendrickson, and Lester B. Lave. 2004. Environmental Management Systems: Informing Organizational Decisions. Paper presented at U.S. Environmental Protection Agency Workshop on Corporate Environmental Behavior and the Effectiveness of Government Interventions. April 2004, Washington, DC.

Mazurek, Janice. 1998. The Use of Voluntary Agreements in the United States: An Initial Survey. Paris: Organization for Economic Cooperation and Development, Environmental Policy Committee. ENV/EPOC/GEEI(98)27/Final. http://www.olis.oecd.org/olis/1998doc.nsf/c16431e1b3f24c0ac12569fa005d1d99/22ab5fb 042b51056c12566d40057e888/$FILE/12E89326.DOC (accessed August 8, 2005).

Metzenbaum, Shelley H. 2001. Information, Environmental Performance, and Environmental Management Systems. In *Regulating From the Inside: Can Environmental Management Systems Achieve Policy Goals?* edited by Cary Coglianese and Jennifer Nash. Washington, DC: Resources for the Future Press, 146–80.

Morag-Levine, Noga. 2003. *Chasing the Wind: Regulating Air Pollution in the Common Law State.* Princeton: Princeton UP.

MSWG (Multi-State Working Group on Environmental Performance). 2004. The External Value Environmental Management System Voluntary Guidance: Gaining Value by Addressing Stakeholder Needs. http://www.mswg.org/documents/guidance04.pdf (accessed August 9, 2005).

Nash, Jennifer, and John R. Ehrenfeld. 2001. Factors that Shape EMS Outcomes in Firms. In *Regulating from the Inside: Can Environmental Management Systems Achieve Policy Goals?* edited by Cary Coglianese and Jennifer Nash. Washington, DC: Resources for the Future Press, 61–81.

Nash, Jennifer, John Ehrenfeld, Jeffrey MacDonaugh-Dummler, and Pascal Thorens. 1999. ISO 14001 and EPA's Region I's Star Track Program: Assessing their Potential as Tools in Environmental Protection. Washington, DC: National Academy of Public Administration. http://www.napawash.org/pc_economy_environment/epafile02.pdf (accessed August 8, 2005).

O'Brien, Tim. 2001. *Ford & ISO 14001: The Synergy between Preserving the Environment and Rewarding Shareholders.* New York: McGraw Hill.

Orts, Eric. 1995. Reflexive Environmental Law. *Northwestern University Law Review* 89: 1227–340.

Parker, Christine. 2002. *The Open Corporation: Effective Self-Regulation and Democracy.* Cambridge, UK: Cambridge UP.

PCA (Portland Cement Association). 2004a. *Cement Manufacturing Sustainability Program.* Washington, DC: Portland Cement Association.

———. 2004b. *Cement Industry Adopts Environmental Management Systems.* Washington, DC: Portland Cement Association.

Perrow, Charles. 1984. *Normal Accidents: Living with High-Risk Technologies.* New York: Basic Books.

Portney, Paul R., and Robert N. Stavins (eds.). 2000. *Public Policies for Environmental*

Protection, 2nd ed. Washington, DC: Resources for the Future Press.

Potoski, Matthew, and Aseem Prakash. 2002. Protecting the Environment: Voluntary Regulations in Environmental Governance. *Policy Currents* 11(4): 9–14.

———. 2005a. Covenants with Weak Swords: ISO 14001 and Facilities' Environmental Performance. *Journal of Policy Analysis and Management* 24(4): 745–69.

———. 2005b. Green Clubs and Voluntary Governance: ISO 14001 and Firms' Regulatory Compliance. *American Journal of Political Science* 49(April): 235–48.

Prakash, Aseem. 2001. Why Do Firms Adopt "Beyond-Compliance" Environmental Policies? *Business Strategy and the Environment* 10: 286–99.

Probst, Katherine N., and David M. Konisky, with Robert Hersh, Michael B. Batz, and Katherine D. Walker. 2001. *Superfund's Future: What Will It Cost?* Washington, DC: Resources for the Future Press.

Project on Alternative Regulatory Approaches. 1981. *Performance Standards: A Practical Guide to the Use of Performance Standards as a Regulatory Alternative.* Washington, DC: U.S. Regulatory Council.

Rees, Joseph V. 1988. *Reforming the Workplace: A Study of Self-Regulation in Occupational Safety.* Philadelphia: University of Pennsylvania Press.

Revesz, Richard L., and Richard B. Stewart (eds.). 1995. *Analyzing Superfund: Economics, Science and Law.* Washington, DC: Resources for the Future Press.

Richards, Kenneth R. 2000. Framing Environmental Policy Instrument Choice. *Duke Environmental Law & Policy Forum* 10(2): 221–85.

Rondinelli, Dennis A., and Gyula Vastag. 2000. Global Corporate Environmental Management Practices at Alcoa. *Corporate Environmental Strategy* 7(1): 288–97.

Rose-Ackerman, Susan. 1973. Effluent Charges: A Critique. *Canadian Journal of Economics* 6: 512–28.

Russo, Michael V. 2001. Institutional Changes and Theories of Organizational Strategy: ISO 14001 and Toxic Emissions in the Electronics Industry. Eugene, OR: University of Oregon, Department of Management.

Salzman, James, and Barton H. Thompson, Jr. 2003. *Environmental Law and Policy: Concepts and Insights.* Westbury, NY: Foundation Press.

Schmalensee, Richard, Paul R. Joskow, A. Denny Ellerman, Juan Pable Montero, and Elizabeth M. Bailey. 1998. An Interim Evaluation of Sulfur Dioxide Emissions Trading. *Journal of Economic Perspectives* 12(3): 53–68.

Speir, Jerry. 2001. EMSs and Tiered Regulation: Getting the Deal Right. In *Regulating from the Inside: Can Environmental Management Systems Achieve Policy Goals?* edited by Cary Coglianese and Jennifer Nash. Washington, DC: Resources for the Future Press, 198–221.

Stahr, Elvis J. 1971. Antipollution Policies, Their Nature and Their Impact on Corporate Profits. In *Economics of Pollution: The Charles C. Moskowitz Lectures,* edited by Kenneth E. Boulding. New York: New York UP, 83–105.

Stavins, Robert N. 1991. *Project 88—Round II, Incentives for Action: Designing Market-Based Environmental Strategies.* A Public Policy Study sponsored by Senator Timothy E. Wirth, Colorado, and Senator John Heinz, Pennsylvania. Washington, DC, http://ksghome.harvard.edu/~rstavins/Monographs_&_Reports/Project_88-2.pdf (accessed March 13, 2006).

———. 1998. What Can We Learn from the Grand Policy Experiment? Lessons from SO₂ Allowance Trading. *Journal of Economic Perspectives* 12(3): 69–88.

———. 2003. Market-Based Environmental Policies: What Can We Learn from U.S. Experience (and Related Research)? Regulatory Policy Program Working Paper RPP-2003-07. Cambridge, MA: Center for Business and Government, John F. Kennedy School of Government, Harvard University.

Sunstein, Cass R. 1999. Informational Regulation and Informational Standing: Akins and Beyond. *University of Pennsylvania Law Review* 147: 613–75.

Teubner, Gunther, Lindsay Farmer, and Declan Murphy. 1994. *Environmental Law and Ecological Responsibility: The Concept and Practice of Ecological Self-Organization.* New York: Wiley.

U.S. Congress, Office of Technology Assessment. 1995. *Environmental Policy Tools: A User's Guide.* OTA-ENV-634. Washington, DC: U.S. Government Printing Office.

U.S. EPA. 1990. *Reducing Risk: Setting Priorities and Strategies for Environmental Protection.* U.S. EPA Science Advisory Board, Document No. SAB-EC-90-021. Washington, DC: U.S. EPA.

———. 2002. 40 CFR 262.105. University Laboratories XL Project – Laboratory Environmental Management Standard. Section 262.105, "What must be included in the laboratory environmental management plan?"

———. 2003a. *2003–2008 EPA Strategic Plan: Direction for the Future.* September 20, 2003. Washington, DC: U.S. EPA.

———. 2003b. National Pollutant Discharge Elimination System Permit Regulation and Effluent Limitation Guidelines and Standards for Concentrated Animal Feeding Operations (CAFOs); Final Rule. *Federal Register* 68(29): 7175, February 12.

———. 2003c. Standards for the Use or Disposal of Sewage Sludge; Final Agency Response to the National Research Council Report on Biosolids Applied to Land and the Results of EPA's Review of Existing Sewage Sludge Regulations. *Federal Register* 68(250): 75531, December 31.

———. 2004. EPA's Strategy for Determining the Role of Environmental Management Systems in Regulatory Programs. http://www.epa.gov/ems/docs/EMS_and_the_Reg_Structure_41204F.pdf (accessed August 8, 2005).

———. 2005a. National Environmental Performance Track. http://www.epa.gov/performancetrack/ (accessed August 8, 2005).

———. 2005b. Environmental Management Systems: Case Studies. http://www.epa.gov/ems/resources/casestudies/index.htm (accessed August 8, 2005).

———. 2005c. 2003–2004 State Innovations Grants Competition Results. http://www.epa.gov/innovation/stategrants/03results.htm (accessed August 8, 2005).

———. 2005d. Guidance on the Use of EMSs in Settlements as Injunctive Relief or Supplemental Environmental Projects. http://www.epa.gov/compliance/resources/policies/incentives/ems/emssettlementguidance.pdf (accessed August 8, 2005).

———. 2005e. EPA Sector Programs—Environmental Management Systems. http://www.epa.gov/sectors/ems.html (accessed August 9, 2005).

Viscusi, W. Kip. 1983. *Risk by Choice: Regulating Health and Safety in the Workplace.* Cambridge, MA: Harvard UP.

Wagner, Wendy E. 2000. Innovations in Environmental Policy: The Triumph of

Technology-Based Standards. *University of Illinois Law Review* 2000: 83–113.

Wilson, Robert C. 2001. Ford Spreads the Word about Its EMS Success. *Pollution Engineering* 33(6): 32–3.

Yosie, Terry F. 2003. Responsible Care Above and Beyond. *Chemistry Business* 31(3): 7.

2

Environmental Management Style and Corporate Environmental Performance

Robert A. Kagan

Few observers doubt that corporate environmental management plays a sig-nificant role in shaping the responses of business firms to regulation and in advancing techniques of environmental protection. For the most part, however, the literature on corporate environmental management remains exhortative and anecdotal rather than empirically grounded and systematic. Hence, despite some serious research on the topic (Prakash 2000; Reinhardt 2000; Coglianese and Nash 2001; Amaral et al. 2000), we still have only limited understanding of why some firms have committed, competent environmental management but others do not. We also do not know very much about how government can best foster high-quality environmental management, nor do we know what we should *not* expect from management-based strategies alone (as compared with what we can expect from traditional, directive government regulation).

This chapter reports on an empirical study of 14 pulp mills in Australia, New Zealand, British Columbia, and the U.S. states of Georgia and Washing-ton.[1] The study sought to identify the relative importance of corporate environmental management itself—as opposed to the effects of regulatory, social, and economic pressures on the mills—in explaining the variation observed in the mills' environmental performance. As it turned out, differences in corporate "environmental management style" correlated with variation in mills' environmental performance. On the other hand, economic constraints seemed to deter even the most environmentally committed managers from leaping substantially ahead of their competitors in environmental control tech-nology or performance. Thus, over time, the largest gains in environmental performance in the industry have stemmed not from enlightened corporate management *per se* but from periodic tightening of environmental standards mandated by governments.

The Place of Management in
Corporate Environmental Behavior

In much economic and legal analysis of regulation, business corporations are viewed as "amoral calculators" that invest in costly environmental protection measures only when specifically required to do so by law, and then only to the extent that noncompliance is likely to be detected and harshly penalized (Becker 1968; Kagan and Scholz 1984). From this standpoint, corporate environmental management, at best, serves simply as a conduit to the company for information about current and forthcoming legal requirements and as an internal policing device, ensuring that operatives comply well enough to keep the company out of legal trouble.

More recently, however, some analysts have viewed corporate environmental management as a vehicle for innovative approaches to pollution prevention and product cycle management that go well beyond mere compliance with law. Corporate managers, it is argued, often have reduced corporate environmental impacts more fully and efficiently than the across-the-board regulations enacted by regulators who necessarily lack detailed knowledge of particular industrial and ecological contexts. For example, following the 1986 federal mandate to create a Toxics Release Inventory (TRI), which requires U.S. firms simply to file public reports listing their emissions, many companies sharply reduced those emissions, although the reduction itself was not required by the TRI law (Konar and Cohen 1997). Some studies indicate that multinational corporations often install in their developing-world facilities the same or similar pollution control equipment as they have in their plants in the United States or Europe, where regulations are enforced much more strictly and effectively (Fowler 1995).

Why would profit-maximizing companies go "beyond compliance" with legal requirements? One common explanation is that as customers, investors, and governments have become more attentive to environmental values, corporate managers have come to believe it is simply "good business" to develop and protect a reputation for being good environmental citizens. Thus, almost half the U.S. manufacturing firms surveyed by Florida and Davison (2001) had undertaken programs to reduce local environmental impacts such as odor and dust, which generally are not closely regulated by government. Michael Porter (1991) has argued that in a highly regulated world, innovative companies can acquire a competitive advantage—gaining new markets or cutting costs—by developing novel methods of reducing environmental problems. Finally, some corporate officials, such as those who love hiking or fishing, may simply care more deeply about the integrity of ecosystems and feel obligated to make their companies as "green" as economically feasible.

One indicator of these shifts in public and corporate attitudes is the recent growth in formal corporate environmental management systems. By the 1980s, most large firms with significant environmental impacts had built substantial environmental management staffs, established internal compliance-auditing systems, and embraced pollution prevention strategies to help meet regulatory demands more efficiently. The International Organization for Standardization's 14000-series standard for voluntary corporate environmental management systems requires firms to go beyond legal compliance by instituting systematic planning for "continuous improvement" in their facilities' environmental performance.

Yet it is not clear to what extent corporate environmental management, in itself, actually produces *significant* improvements in environmental performance, particularly when "beyond compliance" investments are very costly investments and have no clear economic payoff. The "win-win" investments (good for the environment *and* the company's bottom line) highlighted by Porter may not be as prevalent as one might hope (Walley and Whitehead 1994; Palmer et al. 1995; Boyd 1998), or they may arise in some industries and not in others, or may be applicable to measures such as energy conservation and solid waste reduction much more than to other kinds of environmental issues. There is growing (albeit mixed) evidence that firms with formal environmental management systems, such as ISO 14000 plans, have better environmental records (Andrews et al. 2006; Andrews et al. 2003; Florida and Davison 2001; Hendrickson et al. 2004; Russo 2000). Conceivably, however, the firms that adopted those systems faced greater legal threats (which would make regulation and liability law the primary drivers), or greater pressures from local communities and environmental activists (which would make social pressures the primary drivers). For policy purposes, we need to know more about the relationships among regulation, social pressures, economic pressures, and corporate environmental management, and we need to examine those relationships in a variety of contexts.

The "License to Operate" Model of Corporate Environmental Behavior

In the study discussed in this chapter, corporate managers were viewed (and viewed themselves) as operating within the bounds of a multiplex "license to operate." One strand of the broader license is the legal one, embodied in governmentally enforced regulatory permits, other statutory obligations, and liability law. Another strand is each firm's "social license," which reflects the demands of environmental activists, local community groups, and, on occasion, the general public. This social license is enforced by the threat of adverse publicity or complaints to local government and regulators. For example, neighbors who have to live with a pulp mill's unpleasant aroma often demand stricter control of odorous emissions than the facility's legal license demands. The third strand of the license includes the demands of top management, lenders, and investors for cost-cutting and profitability. This "economic license" can limit environmental investment, although markets sometimes punish firms that attract adverse publicity as a result of regulatory enforcement actions or the activities of social license enforcers—such as consumer boycotts. Putting these strands together, the license model suggests that a particular facility's environmental performance is shaped by the relative "tightness" of its regulatory, social, and economic licenses to operate, as enforced by external stakeholders.

The precise terms of each strand of the license to operate, however, often are ambiguous and subject to interpretation and renegotiation by corporate executives. For example, the impact of potential community group action depends in large part on how company management chooses to respond. Similarly, deciding whether a particular environmental investment is a win-win or a "win-lose" investment often is a matter of judgment. Thus corporate environmental management style can be viewed as an intervening variable between external license pressures and corporate environmental performance.

Research Method

Two colleagues and I have undertaken a study of environmental management and performance in 14 pulp manufacturing mills in British Columbia, Canada; Australia; New Zealand; and the states of Washington and Georgia in the United States (Guningham et al. 2003). All four countries have substantial timber and pulp and paper industries. Pulp mills represent a chemical-intensive, heavily polluting sector of the economy. Their effects on local environments have triggered close scrutiny by regulatory agencies and by environmental and community groups. The 14 mills each use a similar chemical-intensive technology, which facilitated meaningful comparisons of their environmental performance. The sample, although not randomly chosen, is reasonably representative of the pulp and paper industry.[2]

We obtained detailed environmental performance data for each mill in 1998–1999. We gathered qualitative data from lengthy on-site, semi-structured interviews with environmental managers at each facility, and in most cases with mill managers and corporate-level environmental managers as well. At each facility, we asked managers which specific kinds of pollution control and prevention technologies were used, guided by a checklist based on the latest EPA regulations and consultants' accounts of industry "best practices." By questioning company officials, we constructed detailed histories of particular environmental, legal, social, and political challenges each facility had experienced and of their responses to these challenges. In assembling these case histories, we also interviewed relevant regulators and local environmental activists.

Quantitative data focused primarily on measures of water pollutants, which are historically regarded as one of the most important aspects of a mill's environmental performance. We were not able to obtain exactly the same environmental performance measures for all mills in our sample over the same time periods, largely because governmental reporting and record-keeping requirements vary somewhat across jurisdictions. But for most mills in our sample we obtained data on the following measures:

- Biological Oxygen Demand (BOD), measured in kg/day, is a standard measure of the organic pollutant content of water, and is a universally important measure of effluent quality. We were able to obtain BOD data for 12 mills in the United States, Canada, and New Zealand for 1998 or 1999 (or both).[3]
- Total Suspended Solids (TSS), measured in kg/day, is the standard measure of the particulate content of water, and is another universally important measure of effluent quality. We were able to obtain TSS data for 12 mills in the United States, Canada, and New Zealand for 1998 or 1999 (or both).
- AOX measured in kg/ton of pulp produced, measures the level of absorbable organic halides (including chlorinated organics such as dioxin) in mills' effluent waters. AOX is used as a proxy measure for dioxins and furans, a family of persistent chlorinated organic compounds that accumulate in the food chain and have been associated with the poisoning of aquatic life, ecosystem damage, and possible human health effects. We were able to obtain comparable 1998 or 1999 (or both) AOX data for 9 mills in the United States, Canada, and New Zealand (but not Australia).
- Chemical Spills. For seven mills in the United States, we were able to obtain data on the incidence of accidental spills of chemicals used in the pulping and bleaching process. Such spills can result in toxic water pollution and overwhelm the mills' waste-water treatment systems; they are also an indicator of the relative quality of the mills' environmental management program.

Environmental Management Style

Formal corporate environmental management systems can be instituted for a variety of reasons, some of them primarily to send an appropriate symbolic message to environmentalists or certain customers. A key factor, Coglianese and Nash (2001) have argued, is managerial "commitment" to continuous improvement in corporate environmental performance. In that spirit, in our study (Gunningham et al. 2003) we used our field research and interviews to identify a multidimensional "environmental management style" for each of the mills and corporations in our sample, incorporating a combination of managerial attitudes and implementing actions.

The focus on both attitudes and actions reflected our sense that managerial attitudes could most confidently be characterized by referring not only to managers' expressed positions on environmental and regulatory issues, but also to their accounts of how their facility had responded to the particular environmental, regulatory, economic, and social challenges that had arisen in their mill's recent history; and to their explanations for the concrete actions they had taken.[4]

To classify each firm's environmental management style, we relied on the interviews to score it on four dimensions of commitment to environmental values:

- firm managers' *environmental ethos,* including such matters as the extent to which they considered environmental improvements to be not only environmentally desirable but economically beneficial to the firm, and the extent to which their explanations focused on environmental goals versus legal compliance;

- managers' degree of responsiveness to environmental demands from regulators, customers, neighbors, and environmental activists (including level of transparency in sharing facility-level information);[5] and

- the apparent *assiduousness* of managers in implementing routines to ensure high levels of environmental consciousness and performance (including such activities as self-auditing, employee training, and close *integration* of environmental and production-oriented decisionmaking).

Based on this information, we classified the 14 facilities in terms of five "ideal types" of environmental management strategies: (1) Environmental Laggards, (2) Reluctant Compliers, (3) Committed Compliers, (4) Environmental Strategists, and (5) True Believers. We developed our overall assessment of each mill's environmental management style before analyzing its effluent data, so that our management measure was independent of the performance measures.

Each successive managerial "type" displayed incrementally greater commitment to compliance (or "overcompliance") with regulatory requirements. The True Believers and Environmental Strategists scanned more intensely and more broadly for environmental information and win-win opportunities, and they were more likely to see an environmental investment as win-win even if it did not clearly meet numeric return-on-investment criteria. That is, they tended to see the pursuit of environmental excellence as a business strategy, not merely a regulation-based constraint. In addition, each successive type displayed a higher level of responsiveness to legal and social stakeholder demands. The True Believers, for example, made their operations more transparent than did the Environmental Strategists. Finally, each successive type manifested a greater commitment to developing reliable implementing routines for their environmental policies, integrating environmental control more tightly with production and quality control.

True Believers appeared to be morally driven in their pursuit of environmental excellence. Environmental Strategists, too, saw aggressive corporate environmental policy as a mode of surviving and competing, but their commitment to environmental excellence appeared to be more contingent on the current and anticipated sociopolitical climate. Committed Compliers tended to define all environmental actions and goals wholly in terms of regulatory requirements, whereas Environmental Strategists (and True Believers) had broader goals and policies. All of these groups took compliance with the law for granted for either moral or strategic reasons. Reluctant Compliers, however, tended to comply only because they did not want to get caught and face regulatory action and they feared the adverse publicity that might flow from such action. In our sample of 14 mills, we identified no Environmental Laggards, two Reluctant Compliers, four Committed Compliers, four Environmental Strategists, and two True Believers.

Mill-Level Environmental Performance and Its Correlates

We observed both similarities and differences in the environmental performance of the mills we looked at in our study (Gunningham et al. 2003). All had taken steps to improve their performance over the past several decades. We also noted important differences among the mills with respect to managers' actions to improve environmental performance beyond legal requirements.

Convergence in Environmental Performance across Mills

Evaluations by government bodies and industry associations, along with our data, all point to a dramatic reduction in the polluting emissions of pulp mills over the last three decades, on the order of 80 or 90 percent for several leading measures of water pollution in wastewater (Armstrong et al. 1998, *123*; AF&PA 2000, *5–12*). There has also been a considerable narrowing of differences between environmental "leaders" and "laggards" in levels of pollution control. None of the mills in this study were regulatory laggards in the sense of being ignorant of or systematic evaders of their "legal licenses." For mills in which there are data over substantial periods of time, BOD and TSS emissions declined absolutely, falling even as production levels remained basically stable.

Beyond Compliance Reductions

Most of the mills in the sample in our study could point to significant instances of environmental improvements that went "beyond compliance." These fell into four categories. The first two of these categories are very closely related to *regulatory* requirements, whereas the second two categories sometimes go "beyond regulation" as well as beyond compliance (and hence may be more attributable to environmental management per se). These categories are:

- *Margin of safety measures* that "overcomply" with regulatory requirements, much as a motorist might drive 5 mph below the speed limit on a well-policed highway. Thus most of the sampled pulp mills had constructed effluent treatment systems that reduced pollution more than regulations or their permits required, or that had larger-than-required capacity, in order to ensure that irregularities or breakdowns in normal operations would not result in a serious violation (and the need to suspend production).

- *Anticipatory compliance measures* that "overcomply" with current legal requirements because a firm anticipates future tightening of the legal license and wishes to avoid the excessive costs of retrofitting relatively new equipment. One manifestation of the "margin of error" and "anticipatory" measures, as indicated by Table 2-1, is that in 1998–99 all of the mills for which we could obtain quantifiable regulatory permit limits had reduced the discharge of key water pollutants to levels well below those specified by their permits.

- *Win-win measures* that both improve environmental performance and increase corporate profits (Porter and van der Linde 1995a, 1995b; Reinhardt 2000), such as new production equipment that is both more efficient and less polluting. For example, during the 1990s, 8 of the 14 pulp mills studied had invested in oxygen-delignification systems, which reduce the need for chlorine or chlorine dioxide as a bleaching agent. Those 8 (in contrast to the other 6) had calculated that oxygen delignification would be cost-effective for their particular operations, reducing expenditures on chemicals. [6]

- *Good citizenship measures* that go beyond existing regulatory requirements and that are not justified in terms of traditional return-on-investment calculations, but rather are justified on the grounds that such actions will enhance the firm's reputation for good environmental citizenship, and will, in the long run, be "good business." Thus several pulp mills had invested in fairly expensive measures to reduce foul odors, primarily to meet social license (rather than legal license) pressures.

Explaining Improvement and Convergence

The improvement and convergence in environmental performance reflected, first of all, increasing stringency and convergence in the firms' legal licenses. The largest reductions in harmful pulp mill effluent have stemmed from investments in expensive technologies, particularly secondary wastewater treatment facilities, oxygen delignification systems, and the substitution of chlorine dioxide for elemental chlorine as a bleaching agent. The investments in these new technologies have been driven by periodic tightening of the performance stan-

TABLE 2-1. Emissions as a Percentage of BOD and TSS Limits in Mills' Regulatory Permits, 1998–1999

	BOD		*TSS*
Facility	*Performance as % of limit*	*Facility*	*Performance as % of limit*
BC2	25	BC2	31
BC3	13	BC3	21
BC4	16	BC4	39
GA1	34	GA1	14
GA3	85	GA3	66
WA1	55	WA1	42
WA2	72	WA2	57
WA3	14	WA3	14
WA4	32	WA4	40

Note: The first two letters of the facility name indicate its jurisdiction. Data are sorted by jurisdiction. BC = British Columbia, GA = Georgia, WA = Washington. No data are available for BC1 and GA2.
Source: Gunningham et al. (2003), 43.

dards in governmental regulations, first requiring sharp reductions in BOD and TSS via secondary treatment, and then later, in the 1990s, mandating drastic reductions in emissions of chlorine compounds (AOX). Expectations of more demanding regulations in the future have been the engine of anticipatory or margin-of-error beyond-compliance investments.

Second, virtually every mill in our study sample, in all four jurisdictions, has experienced an intensification of social license pressures for better environmental performance. These pressures have added to the potency of the regulatory license and surrounded the mills with more monitors, often using regulatory compliance as a benchmark. The social license has pushed many mills to adopt good citizenship measures, such as costly odor-reduction technologies.

Third, the economic strand of the license to operate has become more demanding. An increasingly competitive world pulp market has constrained how far firms can go in a "green" direction. None of the mills in our sample had leapt far ahead of the others by adopting very innovative new environmental technologies or products (for example, by running a totally chlorine free (TCF) bleaching operation, or by creating a completely "closed loop" mill with no discharges to surrounding waterways, or marketing significantly more unbleached paper). The increasingly intense demands of all three strands in the license to operate have meant that a firm can neither afford to drop too low, nor aim too high. Hence, there is considerable convergence in performance.

TABLE 2-2. Environmental Performance by Pulp Mill (Ordered by Performance Level)

BOD		TSS		AOX	
Mill	*kg/day*	*Mill*	*kg/day*	*Mill*	*kg/ton*
BC3	993	BC2	2,349	GA2	0.10
BC4	1,000	BC3	2,484	BC3	0.31
WA4	1,271	WA4	3,147	WA3	0.34
NZ1	1,600	WA3	3,487	BC1	0.46
WA3	1,996	BC4	3,525	NZ2	0.54
BC2	2,302	GA1	3,637	BC2	0.58
GA1	2,367	BC1	4,282	BC4	0.60
BC1	2,549	WA1	5,846	WA4	0.91
WA1	3,848	GA3	7,178	WA2	3.49
GA3	4,663	WA2	7,212	WA1	n.a.
WA2[a]	4,726	NZ1	7,900	NZ1	n.a.
NZ2	4,917	NZ2	8,070	GA1	n.a.
GA2	n.a.	GA2	n.a.	GA3	n.a.
AUS	n.a.	AUS	n.a.	AUS	n.a.

Note: n.a. = not available
a. This facility uses two different pulp production technologies on site, one of which is far more polluting than the technology used at all the other facilities in our sample. The numbers shown here (for BOD and TSS) are figures constructed to estimate what pollutant discharges would have been if all production at this facility had been by the cleaner process. These are not the actual figures discharged by the facility.
Source: Gunningham et al. (2003), 76.

Variation in Environmental Performance

Despite dramatic improvements in environmental performance, pulp mills still generate significant adverse environmental effects.[7] Thus the remaining differences among pulp mills in emissions remain environmentally important. On some key measures—such as BOD, TSS, and AOX—the relative laggards in the sample emitted between three and four times more pollution than leaders (see Table 2-2).

Explaining Variation

Why have some pulp mills done a better job in reducing pollution than others? It is not because of differences in the volume of mill production; effluent levels were not correlated with production level.[8] Nor is it because of differences in the stringency of the mills' various regulatory licenses: the 14 pulp mills' level of environmental performance did not correlate closely with the regulatory jurisdiction in which they operated or the type of regulatory regime they faced.[9] Notwithstanding the supposedly greater deterrent threat of the legalistic American approach to regulation (Harrison 1995), with its strict enforcement and

TABLE 2-3. Management Style and Environmental Performance
(Average Emissions of BOD, TSS, AOX, 1998–1999)

	Management Style						
	True Believer		Environmental Strategist		Committed Complier		Reluctant Complier	
Environmental performance	Value	n	Value	n	Value	n	Value	n
BOD (kg/day)	1,288	3	2,304	4	3,607	4	4,726	1
TSS (kg/day)	4,510	3	3,439	4	6,155	4	7,212	1
AOX (kg/ton)	0.44	3	0.46	3	0.57	2		0

Source: Gunningham et al. (2003), 130.

high penalties for violations, the mills in the United States were as likely to be in the bottom half as in the top half of the environmental performance league. The American mills in Washington (considered by some a politically "greener" state) did not do significantly better on average than those in Georgia. Indeed, variations among mills within each state were as large as differences across jurisdictions.

Moreover, mills' environmental performance was not consistently correlated with the profitability or size (in terms of sales) of the parent corporation.[10] Mills with corporate parents that presumably experienced milder economic constraints in 1998–99 did not have lower BOD, TSS, or AOX levels than mills with corporate parents that were doing less well in that period.[11]

Social license demands did help explain variation in environmental performance. Mills confronted by more active local environmental groups, and those that had been subjected to anti-chlorine campaigns by Greenpeace in the early-to-mid-1990s, tended to have lower pollution emissions and to take more beyond compliance measures such as measures for odor control. Mills that exported pulp to Western Europe, where Greenpeace had generated more concern about chlorine emissions, tended to have somewhat lower AOX emissions than mills that sold pulp to (or were owned by) U.S. paper companies.

In our efforts to explain interfacility differences, the strongest relationship found was between environmental management style and environmental performance. As shown in Table 2-3, average emissions for True Believers were generally lower than those for Environmental Strategists, whose emissions were substantially lower than the average for Committed Compliers, whose emissions were in turn substantially lower than the average for Reluctant Compliers.

These differences relate partly to our finding that, as of 1998–99, True Believers and Environmental Strategists were more likely have invested in state-of-the-art pollution-control technology. In addition, True Believers and Environmental Strategists also achieved larger incremental gains in environmental performance through a more dedicated approach to day-to-day environmental management and policy implementation. In a detailed longitudinal comparison of one True Believer and one Reluctant Complier in the state of Washington, the True Believer showed a pattern of continuous improvement in environmental performance over time, independent of equipment changes, whereas the Reluctant Complier did not (Thornton 2001). Finally, we found that True Believers experienced fewer accidental spills of pulping chemicals, which are both costly and (often) environmentally harmful.

A caveat is in order, however. Environmental management style is far from omnipotent in shaping environmental performance. No mill, not even a True Believer, could ignore the constraints of its economic license. True Believers and Environmental Strategists were not way ahead of others in terms of innovative technologies or approaches. No True Believer or Environmental Strategist had a stellar record of finding and adopting major win-win opportunities that its competitors did not. When a Canadian mill (BC3) stepped ahead of most industry peers by making a large investment in TCF pulp technology, the anticipated demand and "green" price premium did not materialize; BC3, losing money, was forced to retreat from TCF. This experience was perceived in the industry as a negative object lesson.[12]

Conclusion: Management and Regulation

The research described here indicates that corporate management style does matter. The firms we classified as True Believers and Environmental Strategists had tighter controls on their day-to-day operations, thus reducing untoward environmental events. These firms were more aggressive in finding cost-effective ways of meeting and exceeding regulatory requirements. Also the day-to-day incremental change that can stem from committed environmental management can aggregate to significant effects over time. For example, one True Believer achieved an almost 50 percent decline in BOD emissions over a 15-year period, even in the absence of more stringent regulatory rules (Thornton 2001).

At the same time, this research indicates that government regulation and pressure remain crucial factors in driving improvements in environmental performance. After all, almost half the firms in our sample were classified as Reluctant or Committed Compliers. Their behavior, and most of the environmental actions taken by the True Believers and Environmental Strategists as well, were driven by regulatory and social license pressures, current or anticipated. Moreover, the pulp industry, at the time we conducted our research (at the turn of the twenty-first century), was operating within what might be called a "mature regulatory regime." Long a prominent target of regulators, neighbors, and environmental groups, pulp mills have become closely watched. The mills' regulatory licenses are detailed and specific, well enforced, and linked to an often-vocal social license regime. Noncompliance is not an option; a "culture of compliance" prevails (Gunningham et al. 2005). That culture provides the foundation on which competition occurs, reassuring environmentally committed managers that regulatory "bad apples" will not gain a significant competitive advantage by evading regulatory responsibilities. Thus, some level of regulatory enforcement remains necessary, it would seem, to ensure effective corporate environmental management.

A related finding is that in highly competitive, commodity industries such as pulp manufacturing, economic constraints limit how far corporate environmental managers can go. In such industries, firms cannot easily capture and retain market share by developing a reputation for "greenness" (Reinhardt 2000). Even True Believers, we found, shy away from making large expenditures on new environmental technologies unless they are reasonably certain that the level of performance attained by their innovations will become mandatory for their competitors. Hence, in the pulp industry, most large improvements in environmental performance have been linked to expensive new technologies that were mandated, in effect, by periodic instances of tightening all firms' regulatory licenses. The search for continuous improvement by the True Believers and Environmental Strategists was premised on the assumption that the overall trajectory of governmental environmental regulation would continue to be toward more stringent requirements, in which technological innovations by firms in any jurisdiction would sooner or later be reflected in more demanding regulations everywhere.

One implication of this analysis is that regulators would be well advised to identify industry leaders, to find ways to reward them for their beyond-compliance and management commitment efforts, and to work closely with them in determining what innovations are feasible—and hence, what innovations can be made the basis of regulation for the entire industry. There is therefore a proper role both for reflexive regulation, in which businesses are encouraged or required to develop individualized, site-specific plans for reducing their own environmental impacts (Orts 1995; Coglianese and Lazer 2003), and for prescriptive regulation, in which regulators make the leaders' achievements mandatory standards for the industry as a whole.

Notes

1. For a more complete account of the research project and its findings, see Gunningham et al. (2003), Kagan et al. (2003), and Gunningham et al. (2004).

2. The sample included all mills that used the particular production technology in Washington (4), New Zealand (2), and Australia (1). Two of the five mills that used this technology in Georgia were excluded for logistical reasons. In British Columbia, out of 14 relevant mills, we chose one mill with a reputation for environmental per-

formance (based on conversations with regulatory officials and industry consultants); two with a reputation for average or below average performance; and one because its parent corporation operated mills in other jurisdictions (chosen to illuminate the relative importance of corporate culture versus local regulatory and social pressures).

3. The mill-level data on BOD, TSS, and AOX are not easily available to the public in accessible form, amenable to cross-mill comparison. Thus we were able to obtain some data for all 14 mills, but not the same set of data for all mills. Time periods over which the reported emissions data were averaged (daily, monthly, or annually) often varied from jurisdiction to jurisdiction, as did time periods for which various kinds of data were available for different mills (e.g., 30 years, 10 years, 2 years, 1 year).

4. For example, in interviews that typically lasted several hours, we asked mill environmental managers, among others, to describe the five most significant environmental improvements that had been made in their facility in the last decade. For each improvement described, we discussed its genesis, the key issues entailed in deciding to make the improvement, and why the choices made prevailed. We also asked environmental managers for their "wish lists" of improvements they would like to make, and why those improvements hadn't yet been made. We asked environmental managers to rate a number of aspects of their mill's methods for dealing with major sources of air and water pollution in comparison with the methods of other firms in the industry, and to explain why they ranked as they did. We asked for descriptions and assessments of recent interactions with regulatory officials and local and national environmental groups; any legal actions against the facility; and particular adverse incidents, such as chemical spills.

5. By "responsiveness" to demands by regulatory officials and stakeholders, we refer primarily to managers' apparent propensity to accept rather than contest demands or to "give" as little as possible, and to be open and sharing about relevant information.

6. Other mills, particularly those that relied on a different type of wood as an input or that had different customer demands, had declined to install oxygen delignification in the 1990s, since their analyses indicated that the technology would not be cost-effective for their operations. Adopting oxygen delignification, it might be noted, might be considered an instance of anticipatory compliance as well as a win-win measure, since it helped adopting mills achieve reductions in elemental chlorine as a bleaching agent—a goal established by regulators in all the jurisdictions we studied.

7. Pulp mill effluent still decreases dissolved oxygen levels and species diversity in receiving waters (Seegert et al. 1997; Sibley et al. 2001; Kovacs et al. 2002). Although modern production processes and secondary treatment of pulp mill effluent generally prevent acute toxicity in aquatic life, field and laboratory research indicate that pulp mill wastes still cause problems with enzymes involved in normal growth and development in some species (Dube and MacLatchy 2000; Karels and Oikari 2000; Parrott et al. 2000; Chen et al. 2001), as well as changes in population structure, such as age at maturity and ratio of male to female (Munkittrick et al. 1998; Larsson and Forlin 2002).

8. BOD and TSS emissions could be reduced in absolute volume, regardless of shifts in production levels, because the largest reductions came about through upgrades of the mills' end-of-pipe effluent treatment systems, which were built to handle the

largest production volumes.

9. Ranked by BOD emissions, the mills did not cluster tightly by regulatory jurisdiction. Except for AOX emissions for mills in British Columbia, where regulations called for zero AOX emissions by the end of 2002, mills' AOX emissions did not cluster tightly by regulatory jurisdiction. Even with this intense jurisdiction-specific pressure, two of the four BC pulp mills in our sample were slightly above the overall median in AOX emissions.

10. For the most part, facility-level financial data were not available. Corporate-level data are an imperfect proxy, but they might adequately capture the relationships of interest. Although individual mills are expected to be financially independent to a considerable degree, they generally have some level of access to corporate financial resources for major capital investments, particularly for environmental investments where mill-level failings might result in negative reputational consequences for the corporation. Hence, we divided the mills in our sample into three categories based on the average annual sales of their corporate parents during the 1998–99 period. We found *no* significant statistical difference for average 1998–99 BOD, TSS, or AOX emissions for each corporate size category. The same was true when we correlated corporate net income and change in corporate stock price (up or down) with mills' environmental performance.

11. The correlation between corporate size (as measured by annual sales 1998–99) and mill-level emissions was –0.09 for BOD, 0.13 for TSS, and 0.02 for AOX. The correlations between corporate net income (1998–99) and mill-level emissions were 0.21 for BOD, –0.05 for TSS, and 0.46 for AOX, none of which were statistically significant. On the other hand, when we compared contemporary (1998–99) environmental performance data with earlier corporate financial data (1990–94), we found that mills owned by more profitable (defined as the ratio of income to sales) parent corporations in the early 1990s generally had lower emissions in the late 1990s. These mills also had better pollution control technology in the late 1990s.

12. For a more complete account, see Gunningham et al. (2003, *71, 61*). Similar stories are told about two other "leap-ahead" mills by Norberg-Bohm and Rossi (1998, *235*) and Reinhardt (2000, *152-4*).

References

AF&PA (American Forest and Paper Association). 2000. *Environmental, Health and Safety Principles Verification Program, Progress Report.* Washington, DC: American Forest and Paper Association.

Amaral, Deborah, Richard Andrews, Nicole Darnall, Deborah Rigling Gallagher, Suellen Keiner, Eric Feldman, Jessica Jacoby, and Matthew Mitchell. 2000. *National Database on Environmental Management Systems: The Effects of ISO 14001 Environmental Management Systems on the Environmental and Economic Performance of Organizations.* Second Public Report, May 2000. Chapel Hill, NC: University of North Carolina.

Andrews, Richard N. L., Andrew Hutson, and Daniel Edwards Jr. 2006. Environment

Under Pressure: How Do Mandates Affect Performance? In *Leveraging the Private Sector: Management-Based Strategies for Improving Environmental Performance*, edited by Cary Coglianese and Jennifer Nash. Washington, DC: Resources for the Future, Chapter 5.

Andrews, Richard N. L., Deborah Amaral, Nicole Darnall, Deborah Rigling Gallagher, Daniel Edwards Jr., Andrew Hutson, Chiara D'Amore, Lin Sun, Yihua Zhang, Suellen Keiner, Eric Feldman, Dorigan Fried, Jessica Jacoby, Matthew Mitchell, and Kapena Pflum. 2003. Do EMSs Improve Performance? Final Report of the National Database on Environmental Management Systems. Chapel Hill, NC: Department of Public Policy, University of North Carolina at Chapel Hill. http://ndems.cas.unc.edu (accessed November 22, 2004).

Armstrong, Douglas A., Keith M. Bentley, Sergio F. Galeano, Robert J. Olszewski, Gail A. Smith, and Jonathan R. Smith Jr. 1998. The Pulp and Paper Industry. In *The Ecology of Industry: Sectors and Linkages*, edited by Deanna J. Richards and Greg Pearson. Washington, DC: National Academy Press.

Becker, Gary. 1968. Crime and Punishment: An Economic Approach. *Journal of Political Economy* 76 (2): 169–217.

Boyd, James. 1998. Searching for Profit in Pollution Prevention: Case Studies in Corporate Valuation of Environmental Opportunities. Mimeo, Resources for the Future, April 1998.

Chen, C. M., M. C. Liu, M. L. Shih, S. C. Yu, C. C. Yeh, S. T. Lee, T. Y. Yang, and S. J. Hung. 2001. Microsomal Monooxygenase Activity in Tilapia (*Oreochromis mossambicus*) Exposed to a Bleached Kraft Mill Effluent Using Different Exposure Systems. *Chemosphere* 45 (4–5): 581–88.

Coglianese, Cary, and David Lazer. 2003. Management-Based Regulation: Prescribing Private Management to Achieve Public Goals. *Law and Society Review* 37 (4): 691–730.

Coglianese, Cary, and Jennifer Nash. 2001. Environmental Management Systems and the New Policy Agenda. In *Regulating from the Inside: Can Environmental Management Systems Achieve Policy Goals?* edited by Cary Coglianese and Jennifer Nash. Washington, DC: Resources for the Future, 1–25.

Dube, Monique G., and Deborah L. MacLatchy. 2000. Endocrine Responses of *Fundulus heteroclitus* to Effluent From a Bleached-Kraft Pulp Mill Before and After Installation of Reverse Osmosis Treatment of a Waste Stream. *Environmental Toxicology and Chemistry* 19 (11): 2788–96.

Florida, Richard, and Derek Davison. 2001. Why Do Firms Adopt Advanced Environmental Practices (and Do They Make a Difference)? In *Regulating from the Inside: Can Environmental Management Systems Achieve Policy Goals?*, edited by Cary Coglianese and Jennifer Nash. Washington, DC: Resources for the Future, 82–104.

Fowler, Robert. 1995. International Environmental Standards for Transnational Corporations. *Environmental Law* 25: 1–30.

Gunningham, Neil, Robert A. Kagan, and Dorothy Thornton. 2003. *Shades of Green: Business, Regulation and Environment.* Stanford, CA: Stanford University Press.

———. 2004. Social License and Environmental Protection: Why Businesses Go Beyond Compliance. *Law & Social Inquiry* 29: 307–41.

Gunningham, Neil, Dorothy Thornton, and Robert A. Kagan. 2005. Motivating

Management: Corporate Compliance and Environmental Protection. *Law & Policy* 27: 289–316.

Harrison, Kathryn. 1995. Is Cooperation the Answer? Canadian Environmental Enforcement in Comparative Context. *Journal of Policy Analysis & Management* 14: 221–24.

Hendrickson, Chris T., Deanna H. Matthews, and Lester Lave. 2004. Environmental Management Systems: Informing Organizational Decisions. Paper presented at a workshop sponsored by U.S. EPA's National Center for Environmental Economics (NCEE), National Center for Environmental Research (NCER). April 26–27, 2004, Washington, DC.

Kagan, Robert A., Neil Gunningham, and Dorothy Thornton. 2003. Explaining Corporate Environmental Performance: How Does Regulation Matter? *Law & Society Review* 37: 51.

Kagan, Robert A., and John T. Scholz. 1984. The "Criminology of the Corporation" and Regulatory Enforcement Styles. In *Enforcing Regulation*, edited by Keith Hawkins and John M. Thomas. Boston: Kluwer-Nijhoff, 67–95.

Karels, Aarno, and Aimo Oikari. 2000. Effects of Pulp and Paper Mill Effluents on the Reproductive and Physiological Status of Perch (*Perca fluviatilis* L.) and Roach (*Rutilus rutilus* L.) During the Spawning Period. *Annales Zoologici Fennici* 37 (2): 65–77.

Konar, Shameek, and Mark Cohen. 1997. Information as Regulation: The Effect of Community Right-to-Know Laws on Toxic Emissions. *Journal of Environmental Economics and Management* 32: 109.

Kovacs, Tibor G., Pierre H. Martel, and Ron H.Voss. 2002. Assessing the Biological Status of Fish in a River Receiving Pulp and Paper Mill Effluents. *Environmental Pollution* 118 (1): 123–40.

Larsson, D.G. Joakim, and Lars Forlin. 2002. Male-Biased Sex Ratios of Fish Embryos Near a Pulp Mill: Temporary Recovery After a Short-Term Shutdown. *Environmental Health Perspectives* 110 (8): 739–42.

Munkittrick, Kelly R., Mark E. McMaster, Lynda H. McCarthy, Mark R. Servos, and Glen J. Van Der Kraak. 1998. An Overview of Recent Studies on the Potential of Pulp-Mill Effluents to Alter Reproductive Parameters in Fish. *Journal of Toxicology and Environmental Health. Part B-Critical Reviews* 1 (4): 347–71.

Norberg-Bohm, Vicki, and Mark Rossi. 1998. The Power of Incrementalism: Environmental Regulation and Technological Change in Pulp and Paper Bleaching in the US. *Technology Analysis and Strategic Management* 10: 225–41.

Orts, Eric. 1995. Reflexive Environmental Law. *Northwestern University Law Review* 89: 1227–90.

Palmer, Karen, Wallace Oates, and Paul Portney. 1995. Tightening Environmental Standards: The Benefit-Cost or the No-Cost Paradigm? *The Journal of Economic Perspectives* 9 (4): 119–32.

Parrott, Joanne L., Michael R.van den Heuvel, L. Mark Hewitt, Mark A. Baker, and Kelly R. Munkittrick. 2000. Isolation of MFO Inducers from Tissues of White Suckers Caged in Bleached Kraft Mill Effluent *Chemosphere* 41(7): 1083–9.

Porter, Michael. 1991. America's Green Strategy. *Scientific American* (April): 264.

Porter, Michael, and Claas van der Linde. 1995a. Green and Competitive: Ending the Stalemate. *Harvard Business Review* September-October: 120–34.

————. 1995b. Toward a New Conception of the Environment-Competitiveness Relationship. *Journal of Economic Perspectives* 9 (4): 119–32.

Prakash, Asseem. 2000. *Greening the Firm: The Politics of Corporate Environmentalism.* Cambridge, UK: Cambridge UP.

Reinhardt, Forest L. 2000. *Down to Earth: Applying Business Principles to Environmental Management.* Boston: Harvard Business School Press.

Russo, Michael V. 2000. Institutional Change and Theories of Organizational Strategy: ISO 14001 and Toxic Emissions in the Electronics Industry. Paper presented at the annual meeting of the Academy of Management. August 4–9, 2000, Toronto, Ontario.

Seegert, Greg, Derric Brown, and Ed Clem. 1997. Improvements in the Pigeon River Following Modernization of the Champion International Canton Mill. *Biological Sciences Symposium Proceedings.* Atlanta: TAPPI Press.

Sibley, P.K., Dixon, D.G., and Barton, D.R. 2001. Impact of Bleached Kraft Pulp Mill Effluent on the Nearshore Benthic Community of Jackfish Bay, Lake Superior. *Water Quality Research Journal of Canada* 3 (4): 815–33.

Thornton, Dorothy. 2001. *The Effect of Management on the Machinery of Environmental Performance.* Unpublished PhD dissertation, Health Service and Policy Analysis, University of California, Berkeley.

Walley, Noah, and Bradley Whitehead. 1994. It's Not Easy Being Green. *Harvard Business Review* 72 (3): 46–52.

Part II:
Mandates and Regulation

3

Evaluating Management-Based Regulation

A Valuable Tool
in the Regulatory Toolbox?

Lori Snyder Bennear

O ver the last decade, the nature of the risks that government has been asked to control has become more complex, presenting regulators across a variety of policy fields with problems not well suited to traditional policy instruments. A key example of the effect of this complexity on regulation can be found in the financial sector. Financial regulatory systems that functioned well when investors were primarily large institutions proved insufficient when small individual investors entered the market in larger numbers, presenting financial regulators with new challenges.

Although imperfect regulation in the financial sector has most recently captivated the nation's attention, nearly all regulatory agencies face increased complexity, uncertainty, and heterogeneity in the sources of risk they are asked to control, whether these risks are financial, food and drug safety, workplace safety, or environmental quality. In the face of these challenges, it is not surprising that regulatory agencies have increased experimentation with nontraditional regulatory instruments including information disclosure, voluntary initiatives, industry self-regulation, and management-based regulation. However, there is a dearth of empirical research that evaluates how well these innovative regulatory programs actually perform in reducing risk (Bennear and Coglianese 2005; U.S. Government Accountability Office 1998). Without such evaluation it is difficult to ascertain whether these initiatives should be continued and expanded, or the policy experiment failed to deliver the desired results and should be revised or replaced.

This chapter contributes to the debate by evaluating a regulatory innovation called *management-based regulation (MBR)*. Traditionally, regulation has either specified a particular means of achieving a goal or specified the goal and left the means of achieving it up to the regulated entity. In the case of pollution control, technology standards are an example of regulating the means. Performance standards and market-based instruments such as tradeable permits and pollution charges are examples of regulation that mandates the ends to be achieved while giving the regulated entities flexibility to determine the least costly means to achieve those ends. Management-based regulation explicitly imposes neither the means nor the ends. Instead, what is required is that each regulated entity engage in its own review and planning process and develop a set of internal rules and initiatives consistent with achieving the regulation's objectives (Coglianese and Lazer 2003).

MBRs have arisen in at least three policy areas in the United States: food safety, industrial safety, and toxic chemical use and release (Coglianese and Lazer 2003). The first prominent use of management-based regulations in the United States has been in food safety, where the United States has implemented a system known as the *Hazard Analysis and Critical Control Point (HACCP)* for controlling pathogen contamination in the food supply. Food processors are required to use a flow chart to evaluate their production process, identify potential sources of contamination, and evaluate and implement alternatives for reducing contamination risk (U.S. Food and Drug Administration 2001; Coglianese and Lazer 2003).

A second use of MBR is in the area of industrial safety. The U.S. Environmental Protection Agency (EPA) requires manufacturers that use certain chemicals to develop risk management plans. These plans should assess the potential for accidents within their plant and evaluate alternatives for reducing or eliminating this risk (Kleindorfer et al. 2000; Coglianese and Lazer 2003). Kleindorfer describes these planning requirements in detail in Chapter 4.

A final example of MBR in the United States is the regulation of toxic chemical use and release. During the 1990s, 14 states adopted management-based pollution prevention programs. These programs require that plants track the use of regulated toxic chemicals through all stages of their production process, identify alternative production techniques or input mixes that would reduce the use and release of these toxic chemicals, and evaluate each of these alternatives for technical and economic feasibility (Snyder 2004; Coglianese and Lazer 2003).

This chapter examines this last example of management-based regulation: management-based pollution prevention. It focuses on two important sets of questions. First, does MBR actually work? Has the application of MBR to toxic chemical use had an impact on facilities' environmental performance? Is it possible that requiring pollution prevention planning will, by itself, induce private profit-maximizing entities to reduce pollution? Second, assuming that MBR does work, under what circumstances does it work better than other regulatory alternatives? Researchers have often compared different policy instruments in terms of their cost-effectiveness (Hahn 1989; Tietenberg 1990; Stavins 2003). Does MBR attain a given level of risk reduction at lower cost than other policy alternatives, where cost includes both the cost of compliance by the regulated entities and the cost of implementation and enforcement by regulators?[1]

To begin, I explore the question of whether MBR works. In the next section, I develop a general model of how MBR may work and then empirically test it to see whether the use of MBR in state pollution-prevention programs led to improved environmental performance by industrial plants. I specifically examine whether plants in states that adopted management-based regulations for pollution prevention during the 1990s experienced greater decreases in reported releases of toxic chemicals. The evidence suggests that plants in these states did experience more rapid decreases in toxic releases, with releases falling by nearly 30 percent more in states with MBR than in states without MBR. Given the evidence that MBR can be an effective regulatory tool, I then turn to the circumstances under which MBR is likely to be more cost-effective than other regulatory instruments. I develop a model that suggests that MBR is likely to be more cost-effective when:

- there is a large degree of heterogeneity in the population of regulated entities,
- the risk of being regulated is not easily measurable,
- traditional regulations would impose high information burdens on the regulatory agency, and
- there is some uncertainty with regard to the extent of the risk being regulated.

Finally, I offer conclusions and suggestions for future research.

Does Management-Based Regulation Work?

There are three different categories of MBR (Coglianese and Lazer 2003). The first category requires risk-reduction planning but does not explicitly require implementation of the risk-reduction measures identified in the plan. The second category of MBR does not require planning, but does require implementation. The final category of MBR requires both planning and implementation. This chapter focuses exclusively on the first category—namely, regulation that requires planning but does not explicitly require that firms implement their plans. There are two key types of regulations in this category. The first is government-mandated planning for risk reduction. The second is a mandated exchange of information between the regulated entities and the government. In the HACCP program for food safety, the government requires

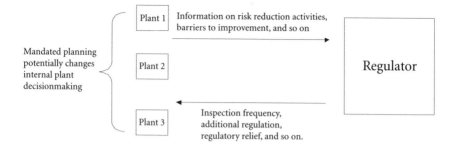

FIGURE 3-1. A Diagrammatic Model of MBR

detailed recordkeeping of all food-related safety issues and requires that these records be made available to inspectors periodically. In the pollution prevention programs, the states require periodic progress reports on pollution reduction activities. The structure of MBR is illustrated in Figure 3-1.

Notice that MBR can affect plant behavior in at least two ways. First, MBR might affect the internal decisionmaking of the plant by revealing opportunities to reduce risk at lower costs or by changing the focus of the key decisionmakers within the organization. I discuss the effect of MBR on internal plant decisions in the section below.

Second, the exchange of information between the plant and the regulator allows the regulator to pool information from multiple plants and better ascertain the nature of the risk, the opportunities for reduction, and directions for further regulatory efforts. This information, if used to direct future regulatory efforts or inspection efforts, can also affect the decisionmaking of the plant. I discuss the effect of the information-sharing component of MBR in a subsequent section of this chapter titled "external interaction with regulators." Because the focus of this book is on private sector environmental performance, the remainder of this chapter is couched in terms of MBR as applied to pollution prevention.

MBR and Internal Plant Decisionmaking

The classic economic model of plant behavior is one in which the plants have limited resources at their disposal and try to allocate those resources to maximize their expected profit or minimize expected costs. Since plant resources include management resources, particularly management time, one can think of an economic model of MBR beginning with a profit-maximizing plant with limited management resources. For simplicity, assume that plants can allocate management resources to investigating pollution-prevention opportunities or investigating other productivity-enhancing improvements such as reducing labor or input requirements. Because the plant is profit maximizing, the plant will optimally choose the allocation of management effort across these two areas in a way that maximizes its expected return. Because different plants may have different expectations of the return from allocating management effort toward pollution prevention, the amount of management effort allocated to this cause will vary from plant to plant.

Even in the absence of regulation that mandates such planning, some plants may optimally choose to engage in some or even a lot of pollution-prevention planning. These plants see higher expected value from allocating management resources to pollution prevention than from allocating these resources to other things. These plants typically expect pollution prevention to enhance profitability, potentially by lowering manufacturing costs or expected regulatory and tort liabilities (Snyder 2004; Florida and Davison 2001). But not all plants will expect to see value in pollution-prevention planning. Alternatively, they may expect that there would be some value in such planning but more value in devoting their management resources to some other area (Florida and Davison 2001). Thus, prior to the establishment of a management-based regulation there is likely to be a nonhomogenous distribution of pollution prevention planning effort across plants. Some plants will already be planning on their own, while others will not be. Among those plants that are engaged in pollution-prevention planning, some will have fully developed environmental management systems and others will have less-structured planning efforts. Other chapters in this volume discuss in some detail the determinants of these differences in nonregulatory driven environmental management styles (most notably Chapter 2 by Kagan).[2]

Management-based regulations mandate that a plant allocate a certain amount of management resources to pollution-prevention planning. The effect of this regulatory mandate on the amount of new planning a plant undertakes will depend on several factors. The first factor is the distribution of pollution-prevention planning effort before regulation. If all plants are already "voluntarily" engaged in highly structured and detailed pollution-prevention planning, then government regulations mandating planning is unlikely to have significant effects on the actual level of pollution. The second factor that determines the effect of MBR is the stringency of the regulation. If the regulation requires only minimal planning efforts, then plant managers may satisfy the regulatory requirements without actually devoting much effort to looking for pollution-prevention opportunities. The final factor to consider is the degree of complementarity between pollution-prevention planning effort and actual reductions in pollution. The definition of *complementarity* is an economic one, namely that increasing consumption of one good increases the value of consumption of the complementary good. The classic economics example of complementarity is peanut butter and jelly. Since many people like to consume peanut butter and jelly together, increasing the consumption of peanut butter makes the consumption of jelly more valuable. In the context of MBR, complementarity between planning and reduced risk implies that increasing planning increases the value of risk reductions, either by decreasing the marginal cost or by increasing the marginal benefits of these reductions. The greater the degree of complementarity between planning and reductions, the greater the expected effect of MBR will be.

To see how complementarity drives the effectiveness of MBR, consider two cases of risk: (1) risk from a terrorist attack, and (2) risk from a lightning strike. In the case of terrorism risk, investments in planning (or learning) about the sources of risk and ways to reduce these risks are likely to increase dramatically the value of any actual risk-reduction activities. For example, after investing in terrorism planning, companies can better allocate resources toward activities that are likely to have a greater effect on actual risk levels (perhaps this would involve devoting more money to screening and security or more money to employee training). This is a case where increased risk planning increases the marginal benefits of risk reduction activities. For lightning strikes, however, it is not clear that investing effort to learning about the sources of risk from lightning strikes is likely to have any real effect on the value of activities to reduce risk from these strikes. Lightning strikes are random; even when the weather conditions are right, nobody can predict precisely where lightning will strike. Investing resources in planning for reduction in risk from lightning strikes is unlikely to uncover ways to reduce this risk at lower marginal cost or higher marginal benefits. The degree of complementarity between risk-reduction planning and risk reductions themselves is likely to be the key determinant of the success of MBR programs.

Four Conceptual Cases of MBR and Pollution Prevention. By combining these three factors one can categorize the effect of MBR at the plant level into four possible cases. In three of these cases, MBR is not likely to be effective at reducing pollution for one of three reasons: because the regulations are weak relative to the amount of pollution-prevention planning that is already occurring, or because the relationship between pollution-prevention planning and cost savings is weak, or because the costs of pollution-prevention activities are extremely high or the benefits are extremely low. In the final case however, MBR is likely to be effective. In this case the government-planning requirements are strong relative to the amount of planning that plants are already doing, and there is a strong complementarity between pollution prevention planning and pollution reductions. I now examine each of these four cases in more detail.

Case 1: Trivial Regulation. If a plant is already engaged in the required level of risk management planning, then the regulation will not require any change in management effort, and hence the regulation should not cause any changes in the plant's environmental performance. For example, if a plant is already ISO

14001 certified, then the plant's environmental management system (EMS) may already include everything the state pollution-prevention regulations require. If this is so, the regulations are trivial to the plant. They may increase the paperwork burden, but they do not fundamentally change the nature of environmental management at the plant. If all plants were Case 1 plants, then one would not expect MBR to make a difference in terms of environmental improvements.

If a plant was not previously allocating the amount of management resources to pollution prevention required by the law, one of three results may occur. The discussion of the next three cases illustrates these results.

Case 2: Weak Complementarity between Planning and Reductions. A Case 2 plant is one for which there is not a strong relationship between increased management effort and discovery of lower costs or higher benefits from pollution-prevention activities. This might be the case if the plant's toxic chemical use is directly tied to its final output. For example, a plant that manufactures vinyl chloride necessarily uses chlorine and either ethylene or acetylene in its production process. All three chemicals are toxic. However, it is not clear how increased management effort will result in the discovery of lower-cost or higher-benefit measures of reducing toxic chemical use for this plant, since these chemicals are fundamental to the production of vinyl chloride. This is a case where there is weak complementarity between management effort and risk reductions. For Case 2 plants one would not expect to see much change in environmental performance as a result of MBR.

Case 3: Strong Complementarity between Planning and Reductions with High Pollution-Reduction Costs. Consider a plant that has high costs of pollution reduction such that costs would have to fall substantially before additional pollution reductions would be profitable. Consider also a plant with low private benefits of pollution reduction such that benefits from decreased tort or regulatory liability would have to increase substantially before pollution reductions would be profitable. For both of these plants, MBR mandates that the plant allocate more management resources to pollution-prevention planning. However, in both cases these additional efforts will result in pollution reductions only if the planning uncovers substantial cost savings or benefits. Thus, despite the additional pollution-prevention planning effort, the plant may not find it profitable to engage in any additional pollution-prevention activities. Absent an unlikely discovery during planning, the plant's environmental performance does not change and the plant is worse off because it allocated management resources away from other potentially profit-enhancing investments. Again, if all plants were Case 3 plants, then one would not expect MBR to result in measurable changes in pollution levels.

Case 4: Strong Complementarity between Planning and Reductions with Lower Pollution-Reduction Costs. In Case 4, consider a plant that is not engaged in the required level of planning before regulation. In contrast to Case 3, this plant has low costs or high private benefits to pollution reduction. In this case, additional management effort directed toward pollution-prevention planning may reveal opportunities for lowering costs or lowering expected regulatory or tort liabilities by engaging in additional pollution-prevention activities. These cost reductions or benefit increases may be sufficient to make these activities profitable. Thus when the regulation forces the plant to increase investment in pollution-prevention planning, the plant chooses to implement pollution-prevention activities that it was not implementing in the absence of such regulation.

It is precisely these Case 4 plants that most regulators have in mind when they argue for pollution-prevention planning as a regulatory instrument. Pollution prevention for Case 4 plants is often mistakenly labeled "win-win" regulation because the environment is improved and profitability of the plant is enhanced. However, the plant is not necessarily better off from a profit standpoint than it was prior to regulation. In order for the regulation to have helped the plant, it has to be the case that the mandated investment in pollution prevention yielded greater cost savings than would have been obtained if the same management resources had been invested in other areas.

To illustrate, imagine a plant that allocates all of its management resources to discovering opportunities to enhance product quality. After MBR some of these management resources must be diverted away from quality improvements to pollution prevention. Further imagine that, after completing a pollution-prevention plan, the plant discovers ways to reduce pollution by recycling some of its inputs and thereby reducing production costs. The firm is clearly better off, right? Unfortunately this is not necessarily true. The key question is: How much would the plant's profits have increased if the plant had stuck to its original allocation of all management resources to quality improvements? It is entirely possible that the plant's profits would have increased still further with its original allocation of management resources.[3] Thus, although Case 4 plants show environmental improvements as a result of MBR, MBR is not necessarily a true win-win policy.[4] This does not mean that MBR is wasteful, only that it has both benefits (improved environmental performance) and costs (reduction in management resources to pursue other profit-enhancing opportunities) just like any other environmental regulation.

Implications of the Four Cases. Management-based regulations that mandate only pollution-prevention planning and do not also mandate subsequent pollution-prevention activities will not necessarily result in measurable improvements in environmental performance through internal changes at the plant. MBR may be ineffective at reducing pollution if the planning requirements are not stringent relative to the amount of pollution-prevention planning that plants are already doing (Case 1). Alternatively, MBR may not be effective if there is a weak relationship between management effort and cost-saving discoveries (Case 2), or if the plants' pollution-prevention activities have very high costs or very low benefits (Case 3). Management-based regulations will improve environmental quality through internal changes at the plant only if (1) there is a strong positive relationship between increasing pollution-prevention planning effort and increasing cost-savings or benefits from pollution reductions, and (2) some plants are not already engaging in the required amount of management effort.[5] That is, management-based regulations will have an overall net effect of reducing pollution only if there are a sufficient number of Case 4 plants in the regulated population.

External Interaction with Regulators

Management-based regulations may also lead plants to improve their environmental performance through interactions between plant managers and regulators. This may be true even if there are few Case 4 plants. In other words, even if the relationship between planning and reductions is weak or the regulation is nonbinding, the information disclosed to regulators through mandated planning might nevertheless have independent effects on facilities' performance.

To see how this information sharing might lead to improved environmental performance, it is helpful to draw from principal-agent theory (Grossman and Hart 1983). In the classic principal-agent model, a principal needs to retain the services of an agent to act on the principal's behalf. The key characteristic of principal-agent relationships is that the principal wants the agent to work toward the principal's objectives, but the agent may not find it in his or her best interest to work toward the principal's goals. For instance, the principal could be an employer and the agent an employee. The employer wants the employee to work diligently eight hours a day. The agent would prefer to shirk if it is possible to do so without the principal knowing. Similarly, the principal could be a homebuyer and the agent a realtor. The homebuyer wants the realtor to disclose any problems with the home. However, the realtor does not receive a commission unless the house sells, a circumstance that provides an incentive to withhold information that may hinder the ability of the home to sell.[6] One can also apply the principal-agent model to a regulatory setting (Laffont and Tirole 1993). In this case the principal is the regulator and the agent is the regulated entity. The principal wants the agent to reduce pollution (or some other measure of risk), but the agent's profits would be higher if pollution was not controlled.

There is a large literature on contracts that resolve the principal-agent dilemma (Grossman and Hart 1983; Hart and Holmstrom 1987). One segment of this literature focuses on the case of multiple agents. The multiple-agent problem is the one most closely related to MBR and is thus worth exploring in more detail.

A classic example of a principal-agent model with multiple agents is a vacuum cleaner dealer that employs several door-to-door salespeople. The dealer cannot observe the individual employee's effort, but can observe each employee's sales. Employee sales are a function of effort and other factors not related to effort, such as the state of the economy, the price of the competition's products, and so forth. Holmstrom (1982) demonstrates that when the factors unrelated to employee effort are common across all employees, then by pooling information on output from multiple agents the principal can actually determine each agent's effort level and reward each employee accordingly.[7] In the vacuum cleaner example, sales may be a function of the general state of the economy, but this factor is common to all salespeople in the same geographic area. Thus, by comparing sales for different salespeople, the dealer can infer which salespeople had the highest effort levels.

An analogous model could apply to MBR. Pollution-reduction levels are a function of pollution-prevention planning effort by the plants, but they are also a function of other factors such as the plants' level of production, other regulatory programs, and so forth. The regulator cannot observe the effort that each plant dedicates to pollution reduction, but can mandate that the plants share information on the level of their pollution reductions. By straightforward application of Holmstrom's findings, if the set of factors that affect pollution reduction levels but are unrelated to each plant's effort are common across plants, the regulator can compare reduction levels across plants to infer which plants engaged in the most pollution-prevention effort.

To use a simplified example, imagine that pollution levels are a function only of pollution-prevention effort and the state of the economy. The regulator cannot observe each plant's pollution prevention effort levels, but since the state of the economy is likely to affect pollution levels similarly at all plants, the regulator simply ranks plants by the magnitude of their pollution levels. This is equivalent to ranking plants based on their level of pollution-prevention effort. Of course, the real world is not this simple. Pollution levels at each plant may be a function of pollution-prevention effort and other factors that are common across all plants (such as the state of the economy), but also plant-specific factors that have nothing to do with pollution-prevention effort (such as the size of the plant). However, even if Holmstrom's findings do not strictly apply, there is still a possibility that mandating that plants share information about their pollution-prevention activities with regulators could lead to a Holmstrom-like result. For example, if most facilities in an industry report rapid decreases in pollution and a few facilities report flat or increasing pollution levels, the regulator may then target these facilities for further investigation. It may turn out that these facilities are fully engaged in pollution prevention planning and their pollution levels are high because of outside factors. But given that the regulator also has limited resources, ranking plants based on their pollution levels can potentially improve regulatory targeting. This kind of ranking can be facilitated by the reporting requirements that are often part of MBR.

Moreover, if information on pollution-prevention activities obtained under MBR is used to determine which facilities should face more frequent inspections, then this may actually increase the value to the plant of engaging in more pollution-prevention activities. In essence, this regulatory targeting acts to increase the benefits of pollution prevention. If the plant knows that by not reducing pollution it may be targeted for more frequent inspections, then the possibility of a reduction in inspection frequency may spur greater pollution-reduction efforts. To make this concrete, return to the example of the plant in Case 3. This facility faced either very high costs or very low benefits of pollution reduction. Just focusing on the internal decisionmaking component of MBR, it appeared unlikely the MBR would result in measurable improvements in environmental performance at a plant like that in Case 3. However, if information on pollution levels were to be used to determine regulatory stringency (like inspection frequency), this would increase the private benefits a Case 3 plant would receive from reducing pollution. This increase in private benefits may be enough to induce the plant to engage in reduction activities that would have proven unprofitable based on a purely internal calculation.

In addition to the information MBR provides to regulators, another way in which MBR may result in improved environmental performance is if plants use these regulations strategically to preempt more costly regulation. This alternative explanation follows a model developed by Maxwell et al. (2000), according to which firms use self-regulation as a strategic tool to preempt government regulation. It is costly for consumers to express their desire for more stringent environmental regulation through the political process. They must write letters to their representatives, engage in protests, make phone calls, or go to the polls and vote. In all of these cases there is a cost of political action (Olson 1971). Plants may be able to preempt this action, and thereby avoid any resulting increase in regulatory stringency, by engaging in some cleanup efforts that otherwise would not be in their best interest. By engaging in pollution-reduction efforts, plants may placate some of the consumers and reduce their desire to bear the costs of political action. If the existence of management-based regulation generally signals a state's willingness to increase regulatory stringency, then plants in those states may have an incentive to demonstrate greater environmental improvement than plants in states without this regulatory threat. According to this model, the resulting change in environmental performance is not a function of the MBR planning requirements per se, but is rather a function of MBR's signal of the state's willingness to impose regulations.

In at least one state, management-based regulation was itself clearly a preemptive regulatory initiative. By the mid-1980s, environmental groups in Mass-achusetts had developed a successful track record of passing state ballot initiatives, and these groups credibly threatened to pursue a new initiative that would impose bans on the use of certain toxic chemicals. Faced with the prospect of very costly bans on toxic chemical use, industry initiated a series of negotiating sessions with environmental groups and state legislators to craft a compromise that could substitute for an outright ban. They succeeded in secur-ing agreement on the Massachusetts Toxic Use Reduction Act of 1989, the first state law in the country to include mandatory pollution-prevention planning requirements (Gomes 1994).

Empirical Evidence

Either because of internal changes within the firm or changes caused by shar-ing information with the regulator, MBR may in theory result in changes in environmental performance. Of course, it is ultimately an empirical question whether management-based regulations actually succeed in promoting improvements in environmental quality. Moving from theory to empirical evi-dence for determining the effectiveness of MBR is not easy. A straightforward comparison of pollution levels in states with MBR and states without MBR might be *suggestive* evidence, but it is hardly convincing proof. There may be other differences among states that explain differences in environmental per-formance. This section discusses how the effects of MBR can be isolated from other potential drivers of changes in pollution levels. Recent empirical work that strives to isolate the causal effect of MBR suggests that MBR has indeed had a positive effect on pollution levels.

Whether MBR is effective at reducing risk can be assessed empirically by iso-lating the *causal effect* of management-based regulation on measures of risk reduction—that is, by identifying the change in performance relative to the change that would have occurred if the management-based regulation had not been implemented (Bennear and Coglianese 2005). In a hypothetical world, one could implement MBR and observe subsequent changes in pollution lev-els. Then, using a time machine, one could go back in time and not implement MBR and observe the changes in pollution levels that occur without MBR. The difference between these two scenarios would then be the causal effect of MBR on pollution. Since we do not have time machines, of course this experiment can never be run. The fundamental problem of causal inference is that we can-not observe environmental performance both with and without MBR for the *same* set of facilities at the *same* time. The research challenge is to use observ-able data to isolate the causal effect of MBR from other explanations of changes in pollution levels. For example, under some circumstances researchers can compare the performance of similar facilities in states with MBR and in states without MBR. In other circumstances, they can compare facilities before and after MBR implementation. And they could also utilize both tools, comparing facilities before and after MBR implementation dates in states with and without MBR regulations (Bennear and Coglianese 2005).

Even if methods exist that can isolate the causal effect of MBR, there are additional complications that stem from the nature of management-based regulations themselves. The first complication occurs when the risks regulated by MBR are low-probability risks. This is true, for example, of management-based regulations aimed at pathogen contamination in the food supply and large-scale industrial accidents. In these cases, measuring changes in risk is difficult. Two facilities, one with a lax risk management plan and one with a rigorous risk management plan, might both experience no incidents in any given period. Yet one still has a greater risk of incidents than the other; it is just that these differences in risk are not measurable as outcomes. In these circumstances, comparing similar facilities across states with and without MBR (and over time) may still not reveal the effect of MBR because these effects are simply not measurable.

An additional difficulty in evaluating the causal effect of MBR is that in regulatory areas such as environmental health and safety, data collection is often built into the regulations themselves. As a result, the researcher observes data on risk measures only for facilities subject to the regulation during years in which the regulation is in effect. For example, researchers have data on toxic releases from the federal Toxics Release Inventory (TRI) program, but only for facilities that are required to report to TRI and only for years in which this regulation has been in effect. Researchers cannot assess the causal effect of the TRI disclosure requirements by comparing releases for facilities that are required to disclose with releases at similar facilities that are not required to disclose. This is because we do not observe releases for facilities that are not required to disclose them under TRI. In general, the marriage of data collection to regulations makes infeasible the use of research designs that compare changes in performance for plants subject to regulation with changes at plants not subject to regulation (Bennear and Coglianese 2005).

Fortunately, state pollution-prevention regulations do not suffer from these particular measurement difficulties. During the 1990s, 14 states adopted management-based pollution prevention laws. These laws targeted reductions in the use and release of toxic chemicals. All states focused on toxic chemicals that require reporting under the federal TRI program; some states also focused on toxic chemicals regulated under the Comprehensive Environmental Response, Compensation, and Liability Act (CERCLA) and the Toxic Substances Control Act (TSCA). The 14 state laws share many common requirements. All states require that plants in the regulated domain track the use of regulated toxic chemicals through all stages of their production process. They also require all such plants to identify alternative production techniques or input mixes that would reduce the use and release of these toxic chemicals, and to evaluate each of these alternatives for technical and economic feasibility. Table 3-1 provides detailed information on the characteristics of these state laws.[8]

In the case of pollution-prevention laws, evaluating the causal effect of MBR is feasible. First, the risk that is being regulated is not a low-probability risk. Pollution is an everyday occurrence at manufacturing facilities. Thus, one can measure changes in environmental performance that result from the regulation. Second, because these laws have been adopted in some states but not others, and measures of pollution exist across all states, the researcher can compare environmental outcomes for regulated and nonregulated facilities. Moreover, these comparisons across states can be made both before and after MBR implementation. The remainder of this section evaluates empirical evidence on the effectiveness of MBR for pollution prevention.

Trends in Toxic Releases. Is there any evidence that MBR has had an effect on pollution levels? Figure 3-2 provides a graph of the trends in toxic releases for plants that are eventually subject to MBR (MBR plants) and facilities that are never subject to MBR (non-MBR plants).[9] The vertical black lines bound the time period in which states enacted MBR. There are several interesting features of these trends in toxic releases over time. First, releases were lower for MBR states before these regulations took effect. This implies that a simple examination of toxic release levels after regulation will overstate the effect of MBR. Some of these differences in toxic release levels existed even prior to MBR implementation and therefore cannot be caused by MBR.

The second interesting feature of the trends in releases over time is that the MBR states appear to have an accelerated rate of decrease in toxic releases compared with non-MBR states. This suggests an effect of MBR on environmental performance, but other differences among states may explain differences in the *rate* of decrease in toxic releases, just as other differences among states must explain the difference in preregulatory *levels* of releases. To isolate the causal effect of MBR, it will be important to control for these other potential determinants of the change in release levels.

The third interesting feature of the trends is that releases in MBR states increase dramatically after 1997. This increase is a function of two plants in Arizona (an MBR state) whose releases increased by an order of magnitude in 1997 and 1998, respectively. Eliminating these two outliers also eliminates the upward trend in MBR state releases. These two Arizona facilities are included to highlight the importance of examining the underlying data rather than simply relying on trends in averages across plants. The TRI data are particularly vulnerable to measurement error because they are self-reported estimates from facilities (Snyder 2004; Graham and Miller 2001).

The trends in MBR and non-MBR releases over time suggest that MBR may have accelerated the rate of decline in toxic chemical releases, but they also

TABLE 3-1. State Pollution Prevention Programs

State	Types of facilities required to plan	Toxic use reporting	Progress reports	Plan is public	Program implementation date	Planning date
Arizona	• TRI reporters • LQGs • Users of more than 10,000 lbs of a toxic chemical (even in a facility that is not a TRI reporter) • Voluntary participants	No	Annual	Yes, reviewed by state officials	1991	1992
California	• LQGs	No	Every four years	Yes, reviewed by third-party auditor	1989	1991
Georgia	• LQGs	No	Biennial	No	1990	1992
Maine	• TRI reporters • LQGs • SARA Section 312 reporters	No	Biennial	No	1990	1993
Massachusetts	• TRI reporters	Yes	Annual	Yes, reviewed by third-party auditor	1990	1994
Minnesota	• TRI reporters with non-zero releases • LQGs	No	Annual	No	1990	1991 for SIC 35-39, 1992 for all others
Mississippi	• TRI reporters • LQGs	No	Annual	No	1990	1992
New Jersey	• TRI reporters	Yes	Annual	No	1991	1992
New York	• LQGs	No	Annual	Yes, reviewed by state officials	1990	Phase in beginning in 1991

TABLE 3-1 (continued). State Pollution Prevention Programs

State	Types of facilities required to plan	Toxic use reporting	Progress reports	Plan is public	Program implementation date	Planning date
Oregon	• TRI reporters • LQGs • SQGs	Yes	Annual	No	1989	1991
Tennessee	• LQGs • SQGs	No	Annual	No	1992	1992 for LQGs, 1994 for SQGs
Texas	• TRI reporters • LQGs • SQGs	No	Annual	No	1991	Phase in beginning in 1993
Vermont	• TRI reporters with non-zero releases • LQGs • SQGs	Yes	Annual	No	1991	Phase in LQGs in 1992, SQGs in 1993, TRI in 1996
Washington	• TRI Reporters • LQGs	No	Annual	No	1988	1992

Note: LQG = Large Quantity Generators; SARA = Superfund Amendments and Reauthorization Act; SIC = Standard Industrial Classification; SQG = Small Quantity Generators; TRI = Toxics Release Inventory.

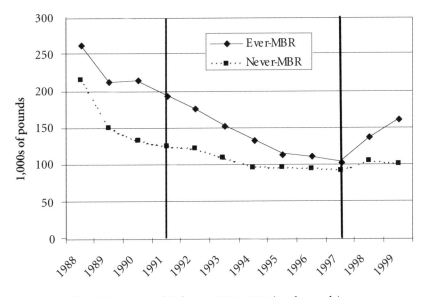

FIGURE 3-2. Trends in Chemical Releases, 1988–1999 (total pounds)

provide reasons to be cautious of this interpretation. Variations in release levels prior to regulation suggest that other factors—such as differences in industry composition or differences in production levels—may explain differences in MBR and non-MBR states. The potential existence of outliers and the extreme effect these can have on average release levels highlights the importance of examining facility-level data rather than relying solely on averages. How does one sort through the data to determine if there is a causal effect of MBR on toxic release levels?

There are two types of empirical evaluations that have been conducted to investigate the degree to which differences in release levels are due to MBR. The first type consists of state-specific program evaluations, which have generally been commissioned by state environmental agencies. The second type compares differences in trends among states with MBR and states without MBR. These comparisons can be done either at the aggregate state level or at the facility level. In principle, these comparative studies are better able to assess the causal effect of MBR. The evidence implies that MBR has had a significant causal effect on toxic chemical releases.

State-Specific Studies. At least two states have conducted evaluations of their MBR laws. In a program evaluation conducted in 1997, three years after its MBR went into effect, Massachusetts reported that it found support for the success of its overall toxic pollution prevention legislation. The state evaluated its MBR law by surveying 645 plants, garnering a response rate of 67 percent (434 responses).

The study found that the vast majority of plants surveyed (81 percent) intended to implement at least some of the source-reduction improvements identified in their mandated pollution-prevention plans. The survey also found that most plant managers stated they would continue to use the planning process even if the legislation were removed. However, the survey did find differential results across plants of different sizes, with smaller plants realizing fewer benefits than larger plants from the program (Keenan et al. 1997).

New Jersey also conducted an official review of its pollution prevention law, in 1996, four years after its law took effect. Consultants for the State of New Jersey visited 115 plants and analyzed data from 405 planning summaries submitted by the facilities to the state. They found evidence that planning was beneficial to plants, with the cost savings associated with pollution-prevention activities outweighing the cost of planning. However, they also found that these results were not uniform across facilities, with smaller facilities receiving fewer benefits (Natan et al. 1996).

Although both of these studies provide support for MBR, they each focus on a single state, making it difficult to separate the effect of MBR from other features of the state regulatory regime. For example, states often have control over permitting and enforcement for many federal statutes including the Clean Air Act, the Clean Water Act, and the TSCA. Could differences in the ways that Massachusetts and New Jersey implemented and enforced these other regulations of toxic chemical releases be responsible for some or all of the observed change in facility behavior? The next section discusses comparative empirical studies that were designed to isolate the effects of MBR from the effects of other influences.

Cross-State Comparisons. Tenney (2000) compared the changes in production-related toxic releases and wastes for fourteen states, seven of which had mandatory pollution prevention planning laws and six of which had voluntary pollution prevention programs. [10] She found that, after adjusting for changes in production, states with mandatory requirements showed greater progress in reducing toxic releases and hazardous waste than states with voluntary programs. She found that production-related waste (that is, waste levels adjusted by production levels) fell by 22 percent in states with mandatory pollution prevention programs, but only by 7 percent in states with voluntary pollution prevention programs. Similarly, she found that releases (again, adjusted for production levels) fell by 51 percent in mandatory pollution prevention states but only by 25 percent in voluntary pollution prevention states.

Unfortunately, the Tenney study was limited by the data available from several states. Moreover, several states that were selected for that study had pollution-prevention planning laws that applied only to a subset of plants that report TRI data. In particular, California, Georgia, and Tennessee have programs that require planning only for large-quantity hazardous waste generators, a subset of all the facilities reporting releases under the TRI program. Large-quantity generators of hazardous waste are facilities that generate more than 2,200 pounds of hazardous waste per month. Many facilities that are required to report releases under TRI are not large-quantity generators of hazardous waste. Thus, a comparison of decreases in TRI releases and waste for Georgia with TRI releases and waste for Alabama may understate the effects of MBR because not all TRI facilities in Georgia are covered by the MBR law. The same holds true for the other states whose laws apply only to large-quantity generators.

In addition, the Tenney study does not correct for other differences across states that might be affecting facilities' toxic releases or hazardous wastes. Differences in industrial composition, environmental preferences of state residents, or other regulatory or economic characteristics of the state might all partially explain why pollution levels vary across states. To isolate the effects of MBR from these other factors, it is necessary to control for other differences across states.

I have sought to overcome the limitations of the Tenney study by comparing changes in toxic chemical releases at plants subject to toxic pollution prevention planning laws with similar changes at plants not subject to such laws (Snyder 2004). My approach was to use a "differences-in-differences" estimator that recognizes that plants are likely to have different environmental performance levels even in the absence of the state regulations. By using such an estimation strategy, I could control for these differences in order to estimate the precise effects of the management-based regulations. These effects were measured by using regression techniques to compute the difference between (1) the average difference between preregulation and postregulation outcomes for the facilities subject to MBR, and (2) the average difference between pre- and postregulation outcomes for facilities that were not subject to MBR, controlling for other facility and state characteristics that might explain both the level and the trend in toxic releases.

To understand how the effect of MBR is isolated from other factors that can vary across states, imagine two facilities, one that is subject to management-based regulation and the other that is not. Further, imagine that these two facilities do not have identical indicators of environmental performance before the regulation takes effect. This is depicted graphically in Figure 3-3. Notice that the facility that is eventually subject to MBR has better environmental performance (lower pollution levels) even before the regulation takes effect. It is clear from the figure that labeling the difference in environmental performance that occurs after the regulation as the causal effect of the regulation would be inaccurate, since some of that difference existed before regulation. The differences-in-differences estimator assumes that, in the absence of treatment, the rate of change in environmental performance would have been the same between the two facilities—that is, the slope of the lines would have been the

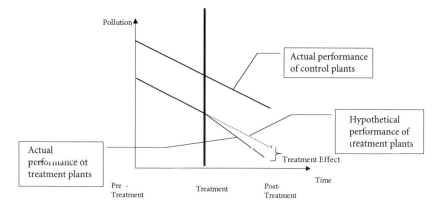

FIGURE 3-3. Differences-in-Differences Estimator

same. The causal effect of the regulation is then just the incremental decrease in pollution in the postregulation period, which is labeled "treatment effect" in Figure 3-3.

It is possible that the trends are not parallel even in the absence of regulation—that is, plants that are eventually regulated may have systematically different trends in environmental performance than plants that are never regulated for reasons other than differences in regulatory stringency. For example, industries may have different rates of technological change. If technological change improves environmental performance, differences in industrial composition among states may lead to differences in trends in toxic releases. To make this explicit, consider two industries: electronics and metal plating. Imagine that there is greater technological change in electronics manufacturing than in metal plating. Further assume that Massachusetts (an MBR state) has many electronics plants and few metal-plating plants, while Ohio (a non-MBR state) has few electronics plants and many metal-plating plants. Even in the absence of regulatory differences one might expect to see a faster decline in toxic releases in Massachusetts than in Ohio simply because of the differences in industrial composition.

An examination of the trends before 1991 in Figure 3-2 indicates that, in fact, MBR plants and non-MBR plants may have different preregulatory trends in total releases. As a result, the differences-in-differences estimator could be biased. To correct for any possible bias and thereby ensure that the measured effect is due to the regulation alone, I controlled for other characteristics that also can affect trends in plants' environmental performance. Specifically, I controlled for industry classification, large-quantity generators status, economic and demographic characteristics, and differences in other regulatory regimes to isolate the effect of MBR better. I found that facilities subject to MBR reduced their total pounds of chemical releases by 60,000 pounds more than comparable facilities not subject to MBR (Snyder 2004). I also found the effect of MBR to be statistically significant for the first six years after adopting the regulation, providing evidence that MBR has lasting effects on changes in environmental performance (Snyder 2004).

These results suggest that MBR has had a causal effect in terms of improving facilities' environmental performance. They do not, however, explain why MBR has been successful. Is it because increased planning effort changed the internal decisionmaking of the plant? Or did information sharing between the regulator and the plant lead the plants to improve in order to avoid future regulatory or enforcement actions? Maxwell et al. (2000) have provided some preliminary evidence on this question. Examining changes in TRI emissions of 17 key toxic chemicals from 1988 to 1992 at the aggregate state level adjusted by the value of shipments, they compared states with pollution-prevention legislation in place by 1991 with states without such legislation. They found no statistically significant effect of state legislation on the change in toxic releases of these 17 chemicals.[11] However, they did find greater changes in total releases in those states where residents had higher rates of membership in environmental organizations. Viewing environmental-group membership as a proxy for the preferences of the state residents and the potential threat of future environmental regulation, they argued that firms are more likely to reduce emissions voluntarily if the threat of future regulation is high.

Other studies have provided evidence consistent with both the internal decisionmaking and the information-sharing explanations for MBR's success. New Jersey and Massachusetts found that larger plants were more likely to improve environmental performance as a result of pollution-prevention planning (Natan et al. 1996; Keenan et al. 1997). Consistent with the internal plant dynamics theory, it might be that larger plants have a greater ability to convert planning into positive action, while smaller plants are subject to the planning burden without having the resources to convert their plans into substantive action. However, it might also be the case that larger plants anticipate that future regulatory or inspection actions are more likely to be directed toward them, and hence these larger plants have greater incentives for engaging in voluntary improvements to avoid these future actions. Clearly more research is needed to understand why MBR affects plant performance.

When Is MBR Preferable?

Given the evidence suggesting that pollution-prevention planning can be an effective regulatory tool, the next logical question to ask is: When are management-based approaches the preferable regulatory tool? Risk regulators have a growing set of policy instruments available to them, including bans, taxes, tort liability, technology standards, performance standards, tradable permits, information disclosure, industry self-regulation, and management-based regulation (Hahn 1989; Richards 2000; Stavins 2003). Different sets of these instruments will be appropriate for different types of risk regulation. The challenge for regulators is to determine the conditions under which each type is the appropriate choice.

Determining what is an "appropriate choice" can be based on several criteria against which regulatory instruments can be compared. One can compare instruments' ability to achieve the socially optimal level of pollution control—the *efficiency criterion* (Baumol and Oates 1988; Helfand et al. 2003). One can compare the instruments in terms of their ability to achieve any given level of risk reduction at the lowest cost—the *cost-effectiveness criterion* (Baumol and Oates 1988; Helfand et al. 2003). One can also compare instruments on other criteria, such as how equitably risks and risk reductions are distributed. Similarly, one could compare the effect of different instruments on the policy process itself, assessing whether a particular policy instrument promotes public participation or enhances the relationship between government, the public, and industry.

No single regulatory instrument is appropriate for every environmental problem. The characteristics of the pollutants (i.e., uniformly mixed, global or local, acute or chronic exposure), the characteristics of the regulated entities (i.e., competitive market, degree of heterogeneity in the production process, small or large facilities), and the nature of regulatory institutions (i.e., democratic, cooperative, or confrontational) can and should affect the choice of policy instrument. Table 3-2 highlights some of the key characteristics of risks, regulated entities, and regulatory institutions that might affect the suitability of a given policy instrument.

In the environmental policy area, researchers have gone to great lengths to ascertain how well different policy instruments perform under various conditions. The optimal choice of a policy instrument has been shown to depend, among other things, on the heterogeneity of the regulated population (Newell and Stavins 2003), the uniform or nonuniform nature of the pollutant (Hahn 1989), the nature of the uncertainty surrounding the benefits and costs of pollution control (Weitzman 1974), the existence of other distortionary taxes (Goulder 2000), the nature of technological change (Milliman and Prince 1989; Jung et al. 1996), and the irreversibility of damages (Arrow and Fisher 1974).[12]

Where does management-based regulation fit into the mix of regulatory instruments available for risk reduction? Unfortunately, the literature on instrument choice has largely ignored management-based regulation. One exception is Coglianese and Lazer (2003), who argue that MBR is best suited for policy situations where the risk itself is difficult to measure and the regulated entities are highly heterogeneous. Let us examine these two characteristics in more detail and see why MBR might be a preferable and appropriate regulatory instru-

TABLE 3-2. Key Factors Determining Policy Instrument Choice

Characteristic of risk	Characteristics of risk sources	Characteristics of policy institutions
Measurability of risk	Heterogeneity of sources	Democratic
Determinants of risk—location, demographics, affected population	Market structure	Enforcement resources
Level of uncertainty surrounding risk	Dynamic properties of product or market	Existing tax structures
Asymmetry of information about risk	Uncertainty surrounding cost of risk reduction	Centralized or decentralized

ment when these characteristics are present. In addition, this section highlights two additional characteristics of the risk and the regulated entities that might make MBR a preferable regulatory alternative compared to other instruments: the information burden for regulatory agencies and uncertainty in the nature of the risk being regulated.

Characteristic 1: Measurability of the Regulated Risk

The first characteristic that might make MBR a preferable regulatory instrument concerns the measurability of the risk being regulated. When crafting environmental regulation, regulators have some discretion in selecting the exact measure of environmental performance that will be targeted. They can choose to measure a facility's performance according to

- inputs, such as the sulfur content of coal;
- releases, such as the pounds of dioxin emitted;
- concentration, such as measured levels of biological oxygen demand (BOD) in the water;
- exposure, such as weighted concentrations based on the number and types of people exposed; or
- risk, such as a weighted measure based on concentration, exposure, and toxicity of the pollutant.

In general, these different measures can be arrayed along a spectrum as shown in Figure 3-4.

The measures farther to the left of the figure tend to be easier to quantify; the measures farther to the right often provide a more accurate assessment of the outcomes of greatest interest to environmental regulators. Yet most environ-

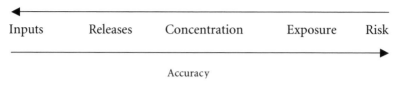

FIGURE 3-4. Tradeoffs in Targets for Environmental Regulation

mental regulation in the United States has targeted inputs and releases, with some regulations focusing on concentration.[13] Few, if any, regulations have specifically set targets for exposure or risk. The chief reason is that there is a tradeoff between accuracy and measurability. If the environmental outcome needs to be measured for regulation to be enforced, then the regulation needs to target inputs, releases, and concentrations, at least until more reliable measures of exposure and risk can be developed.

As Figure 3-4 implies, there is a tradeoff between measurement and enforcement cost and the accuracy of the target as a measure of environmental performance. The extent of that tradeoff depends on the specifics of the risk being regulated. In the case of toxic chemical use, risk is a function of the toxicity of the chemical, its concentration, the duration of exposure, and the characteristics of the exposed population (U.S. EPA 2002b). As a result, there is not a simple relationship between risk and a more readily measurable indicator of environmental performance that regulation could target as a proxy for risk. Regulators can choose to target an imperfect but available proxy for risk, or they can invest in better measures of risk and target those measures.

Management-based regulations can be thought of as regulations that target a single particular type of input called *management effort*. If three conditions apply, one might argue that targeting the regulation at management effort is a reasonable approach to regulating risk from toxic chemical use:

• there must be a strong relationship between management effort at pollution prevention and risk reduction,
• management effort must be able to be measured and monitored, and
• there must be no other measure of risk that can be easily measured.

This framework dovetails with Coglianese and Lazer's (2003) argument that management-based regulations have appeal in areas of risk regulation where other "performance-based" measures of risk are not available to regulate. However, the framework adds two additional restrictions, namely that increasing management effort has to have a measurable and monotonic effect on risk, and that management effort has to be either easier to monitor or a stronger signal of risk than other available indicators of environmental performance.

Characteristic 2: Heterogeneity of Regulated Entities

A second characteristic by which one can classify and compare regulatory instruments is the degree of heterogeneity among regulated entities. The effect of heterogeneity on instrument choice is well studied (Newell and Stavins 2003; Tietenburg 1990). Most of the work on heterogeneity has focused on market-based policy instruments such as pollution taxes and tradable pollution permits. The appeal of market-based instruments on cost-effectiveness grounds is due, in part, to their ability to allocate the distribution of pollution reduction unevenly among regulated entities. When plants have different costs of pollution reduction, uniformity of regulation across plants is costly (Tietenburg 1990). In general, the greater the heterogeneity in the population of regulated entities, the greater the degree of nonuniformity in the regulation needed to be effective. Thus, heterogeneity in the population of regulated entities makes market-based policy instruments more attractive. [14]

Similarly, greater heterogeneity may make MBR more attractive. Management-based regulations will also yield nonuniform changes in environmental performance, since each plant develops its own pollution-prevention plan and voluntarily decides whether to undertake activities that follow from that plan. Plants with lower costs of pollution reduction are likely to engage in more reduction activities, a distribution of reduction activity that is more cost-effective than requiring all plants to engage in the same type of pollution-prevention activity.

Heterogeneity exists in many forms. Typically one thinks of heterogeneity in a cross-section of plants—that is, at any given time different plants are using different production technologies, are of different size, and have different pollution-control costs. However, it is also important to consider heterogeneity over time, or—put differently—the rapidity of technological change in the industry. If the industry is very dynamic, with new production technologies coming on line rapidly, then the regulator will want to avoid locking in a particular technology or even a particular performance standard that might soon be obsolete. One of the advantages of market-based instruments is they provide the correct incentives for plants to continue to adopt pollution-reducing technology (Jaffe et al. 2003). By encouraging regular updating of the toxics use reduction plan, and thus encouraging continued search for new ways of reducing toxics at lower costs, management-based regulations also preserve flexibility over time. Thus, MBR might be considered a more valuable regulatory alternative for industries that are highly dynamic.

Characteristic 3: Information Burden on the Regulator

In addition to measurability of risk and heterogeneity of regulated entities, MBR might be a preferable alternative in the face of a third characteristic—a heavy information burden, which is required to pass more traditional standards. Along the spectrum in Figure 3-4, one could imagine implementing both uniform and nonuniform standards. If part of the appeal of market-based instruments is their ability to provide nonuniformity in the distribution of pollution control, thereby lowering pollution-control costs, the remainder of their appeal is that they do this with a minimum of information required by the regulator (Stavins 2003, *359*). The regulator does not need to know the specific nature of production at each plant to determine how much pollution control each plant will undertake. Once the tax level is set or the quantity of permits is determined and allocated, each plant uses its own information on abatement costs to determine the best course of action. Some plants may choose to clean up more and sell permits, others may choose to clean up less and buy permits, but the government does not need to know in advance which plants will do what.

In regulating toxic chemicals, one could imagine developing detailed management plans, technology requirements, or performance standards that are adapted to the specific nature of production, toxics use, substitute availability, and other plant-level characteristics. However, constructing these nonuniform regulations would require a tremendous amount of information on plant-specific characteristics that is generally not available to regulators (Coglianese and Lazer 2003). Management-based regulations try to harness the informational advantage that plants have about their own production processes and opportunities for environmental improvement, and thereby reduce the informational burden on the regulator.

Characteristic 4: Uncertainty over the Nature of the Regulated Risk

The fourth and final characteristic of settings in which MBR may be a preferable regulatory policy instrument concerns uncertainty over the actual nature of the risk being regulated. When there is uncertainty over the benefits of risk reduction—as, for example, because a chemical is suspected of being toxic but the degree and nature of its toxicity is debatable—then traditional standards may be viewed as committing the firm to a specific regulatory target without sufficient information to determine whether the target is appropriate. In such a case, the regulator may want to be able to learn more about the nature of the risk before committing to a particular regulatory standard. The value of preserving the option of learning about environmental risks is related to the literature on irreversible investments and option value (Arrow and Fisher 1974; Pindyck 1991; Hassett and Metcalf 1995). When investment cannot be easily reversed and the benefits or cost of the investment are uncertain, there is value to waiting for the uncertainty to be resolved. For example, retrofitting older homes with more energy efficient insulation and heating systems is largely irreversible. Once the systems have been installed they are costly to remove. The benefits of making older homes more energy efficient is largely a function of the price of heating oil or natural gas, which is uncertain. Thus, there is often value to waiting to learn more about future oil prices before committing money to install the energy-efficient systems. Hassett and Metcalf (1995) have shown that this option value can largely explain the low rate of conservation investment that resulted from the energy tax credits.

Option value works in the same way for regulatory agencies. Given that traditional notice-and-comment rulemakings are relatively expensive and difficult to alter, regulators may prefer a less rigid standard when uncertainty over the nature of the risk being regulated is great. In these circumstances, MBR might provide a reasonable alternative for three reasons. First, the regulator can achieve some changes in behavior by plants without committing to a specific regulatory standard. Second, the mandatory information-sharing component of MBR helps the regulator learn about the costs and private benefits of risk reduction. And, finally, the regulator has time to learn more about the nature of the risk and determine whether future regulatory efforts are required.

Examples of Desirable MBR Use

Although all four of these characteristics or criteria help explain the circumstances under which MBR might be the most appropriate regulatory instrument, not every problem for which MBR may be considered will necessarily exhibit all four characteristics. Different combinations may lead to MBR being a better regulatory instrument than other feasible alternatives. To see how these four characteristics or criteria might be combined in ways that make MBR a more appropriate regulatory alternative, we examine two different uses of MBR: ensuring food safety and preventing pollution.

In the case of ensuring food safety, three of the four characteristics are met: measurability, heterogeneity, and information burden. Pathogen contamination of the food supply is a difficult risk to measure. Tests are available for key pathogens, but many require so much time to conduct that the food is on the table before the test results are known. Thus setting a standard—for example, requiring that meat not contain more than a certain amount of *E coli*, is not enforceable. But this feature alone does not necessarily suggest MBR is the best regulatory instrument for ensuring food safety: one could still impose uniform "best practice" standards for food handling and processing. However, heterogeneity exists in the population of food processing facilities that might make infeasible a set of best management practices that function uniformly well across a wide range of facilities. Finally, designing a series of best practice standards that apply differentially to different types of facilities could be done, but would place a large information burden on the regulator.

Notice that in the case of ensuring food safety, the uncertainty of the risk was not a key factor. In fact, there is fairly strong evidence linking food contamination to health problems, so, compared with other risks, there is little uncertainty for food safety. But the combination of immeasurability, heterogeneity in the population of regulated entities, and a high information burden on the regulator is arguably sufficient to justify the use of MBR as a regulatory instrument for ensuring food safety.

A different combination of the four characteristics would justify the use of MBR for toxics pollution prevention and control. In the case of toxics pollution, there are measures, albeit imperfect ones, of toxic chemical release levels from each facility. Every year large manufacturing facilities are required publicly to disclose their releases of nearly 600 different toxic chemicals under the federal TRI program. Thus, toxics regulation could require that facilities not release more than a certain number of pounds or require certain percentage reductions in toxic chemicals. The targets of these regulations would be measurable and enforceable. Given that, why would anyone ever argue that MBR should be used to regulate toxic chemicals?

First, there is again a large degree of heterogeneity in the population of facilities that release toxic chemicals. Therefore, uniform standards are likely to be quite costly for facilities, and nonuniform standards place a high information burden on regulators (Stavins 2003; Tietenberg 1990). However, these two conditions are still not sufficient to justify the use of MBR for toxics control. In similar situations for other pollutants, heterogeneity and regulatory information burden have led to the use of market-based regulatory instruments such as tradable pollution permits (Stavins 2003; Tietenberg 1990). Tradable permits still guarantee an overall level of reduction in pollution, but they attain this reduction at the least cost by allowing facilities with higher costs of pollution control to pay for reductions at facilities with lower costs (Stavins 2003; Tietenberg 1990).

The argument for MBR in the context of toxics pollution seems to hinge on uncertainty over the nature of the risk. By definition, all toxic chemicals reportable under the TRI program cause either human health or environmental/ecosystem damage at some level (U.S. Code 1986).[15] But the degree of damage, particularly at current levels of exposure, is not clear for at least some of these chemicals. Mercury and arsenic are two recent examples of how uncertainty over the risk posed by a known toxic chemical impaired the ability to set traditional standards. For example, the January 2004 *Federal Register* notice for a new EPA regulation of mercury releases from utilities stated:

> Exposure to Hg [mercury] and Ni [nickel] at sufficiently high levels is associated with a variety of adverse health effects. The EPA cannot currently quantify whether, and the extent to which, the adverse health effects occur in the populations surrounding these facilities, and the contribution, if any, of the facilities to those problems (U.S. EPA 2004a).

In 1999, the National Research Council (NRC) provided a scientific assessment of the effects of arsenic in drinking water to aid in EPA's decision on whether to lower the standard for arsenic below the established 50 parts per million (ppm) standard. As with mercury, the debate over arsenic standards has not focused on the potential toxicity of the chemical, but rather on the health effects of exposure to this chemical at current levels. The NRC report stated:

> The subcommittee concludes that there is sufficient evidence from human epidemiological studies in Taiwan, Chile, and Argentina that chronic ingestion of inorganic arsenic causes bladder and lung cancer, as well as skin cancer. With minor exceptions, epidemiological studies for cancer are based on populations exposed to arsenic concentrations in drinking water of at least several hundred micrograms per liter. Few data address the degree of cancer risk at lower concentrations of ingested arsenic (NRC 1999).

As the mercury and arsenic examples illustrate, uncertainty over the nature of the risk posed by exposure to known toxic chemicals can limit the ability of regulatory agencies to use more traditional environmental standards. Until more is learned about the underlying science of human exposure to these chemicals, specifying regulatory standards for these chemicals may result in standards that prove either too stringent or not stringent enough. Although all regulations can be changed in the presence of new information in theory, it is often costly and politically difficult to do so. MBR provides a way to achieve some reductions in toxic pollution while learning about the costs and benefits of further regulatory action.

The above discussion has focused on theoretical arguments for establishing occasions when MBR might be a viable regulatory alternative. Ideally, one could use empirical evidence to compare the effectiveness of different policy instruments and determine which instrument is most appropriate for a given class of problems. Of course, to date there has been no empirical evaluation comparing the performance of MBR with that of other regulatory policy instruments. This empirical comparison is difficult to make in large part because different regulatory instruments are often used exclusively in one policy context. Facilities are usually not simultaneously subject to MBR, market-based instruments, and command-and-control instruments for the same pollutant. Even if they were, isolating the effects of each instrument would be quite challenging. Nonetheless, identification of policy areas where these comparisons can be made empirically should be a high priority for future research.[16]

Conclusion

This chapter presents both theoretical and empirical evidence of the effectiveness of MBR, focusing on its use in controling toxic chemical pollution. From a theoretical perspective, the chapter demonstrates that MBR can be effective in two ways. The first is by changing the internal decisionmaking of the plant. MBR forces plants to allocate management resources toward pollution prevention. As a result, some plants discover methods for reducing toxic chemical use or releases at lower costs. The key factors in determining whether MBR will affect the internal decisionmaking of the plant is (1) whether the plant is already engaged in the level of pollution-prevention planning required by regulation, and (2) the strength of the connection between increased management effort and the discovery of lower-cost or higher-benefit methods of reducing toxic chemical use and release. If plants are already engaged in the amount of planning that MBR requires, then government regulation is unlikely to result in measurable changes in facility-level pollution. Similarly, if there is a limited relationship between management effort and pollution reductions, then requiring facilities to devote more management effort to investigating possibilities for reducing chemical use and release will not be fruitful.

The second theoretical factor in determining whether MBR may be effective is the level of information sharing between the regulated entities and the regulator. MBR requires plants to share with the regulator the results of their pollution-planning efforts. This information sharing can provide the regulator with critical information on both the costs and private benefits of toxic chemical use reductions. It can also allow the regulator to compare the performance of plants in the same industry and to allocate government resources to inspecting plants that appear to be laggards. The key determinant of the success of information sharing is the degree to which the regulator can make valid comparisons among plants' environmental reports.

Beyond these theoretical arguments of why MBR might work, new empirical evidence has emerged. This evidence shows that plants subject to MBR have experienced greater reductions in toxic chemical releases than they would have in the absence of these regulatory initiatives. The most rigorous study on the topic has found that facilities subject to these regulations reduced toxic chemicals by nearly 60,000 pounds more than comparable facilities not subject to the regulations. This empirical work, comparing the relative performance of facilities in states with and without MBR, complements prior empirical work that investigated facility responses to specific states' programs. In general, the empirical evidence supports the conclusion that MBR is effective for toxic chemical reduction.

Given that MBR can be effective, under what circumstances might MBR be a better regulatory instrument than other alternatives? Four characteristics that would indicate such circumstances are

- the measurability of risk,
- the degree of heterogeneity of the regulated entities,
- the degree of information required to promulgate traditional regulations, and

- the degree of uncertainty surrounding the nature of the risk being regulated.

Two examples of the current use of MBR—ensuring food safety and preventing pollution—were used to demonstrate how these characteristics might be combined in ways that make MBR an attractive regulatory option. In the case of food safety, the risk of bacterial contamination is not easily measurable quickly enough to affect consumption of contaminated food. This has made binding standards on bacteria levels infeasible as an exclusive basis for regulation. In addition, there is a great deal of heterogeneity among food processing plants, thereby limiting the ability to use uniform best management practice standards. Finally, the information burden for regulatory agencies to design facility-specific regulations would be extremely high. Given this combination of characteristics of the regulated entities, the risk being regulated, and the regulatory institutions, MBR seems to be a reasonable policy measure to control food contamination.

A different combination of characteristics seems to justify the use of MBR to prevent pollution. In this case, there are measures of pollution outcomes available at the facility level. Traditional performance standards that limit toxic chemical releases or market-based instruments such as tradable permit systems could be used and enforced. The key characteristic of pollution prevention that makes MBR justifiable is that, for many of the chemicals in question, the degree of damages from current levels of exposure is uncertain. It is difficult to set standards in the absence of good information on the cost and benefits of toxic chemical control. MBR provides a way to achieve some reductions in toxics pollution while learning about the costs and benefits of further regulatory action.

Empirical research has shown that MBR can be a viable regulatory strategy. It therefore deserves to be studied by researchers and considered by policymakers as an alternative to other, better-recognized policy instruments. In addition, there are circumstances under which MBR may be a superior regulatory instrument compared with alternatives. Further research will be needed to determine just how appropriate MBR might be in different circumstances, but the evidence from state pollution-prevention planning laws shows that MBR has earned a place in the regulator's toolbox.

Acknowledgments

This research was funded in part by the Science to Achieve Results (STAR) grant program at EPA (grant number R829689) and a grant from the Heinz Scholars for Environmental Research program. This research does not reflect official policy of the U.S. Environmental Protection Agency or the Heinz Family Foundation.

Notes

1. Cost-effectiveness is not the only criteria by which policy instruments should be compared. Other criteria for comparison include whether the policy results in a more equitable distribution of the costs and benefits or changes the policy process in desirable ways.

2. In Chapter 2, Kagan finds that some firms—what he calls the "True Believers"—always overcomply. Other firms may be strategic overcompliers, while still others may only be reluctant compliers. Similarly, these management styles are likely to be highly correlated with the firms' allocation of management effort toward pollution prevention.

3. The plant could be better off with MBR than without MBR if some management resources were idle and those resources were devoted productively toward pollution prevention, or if managers got lucky in their pollution-prevention planning—for example, if managers discovered greater-than-anticipated cost savings from pollution reduction.

4. Porter and van der Linde (1995) argue that regulation might increase competitiveness by focusing firms' attention to improvements in environmental performance that also enhance economic performance, so called win-win opportunities. Palmer et al. (1995) provide the neoclassical economics counterpoint for why situations that ex post may appear to improve both environmental and economic performance are not necessarily win-win situations ex ante.

5. Snyder (2004) provides a proof of this assertion.

6. This incentive problem is the primary rationale for regulations that require disclosure of "material facts" by real estate agents.

7. This is the foundation for relative performance standards—standards that reward or punish based on relative, rather than absolute, performance. For more information on relative performance standards and their use to solve principal-agent problems, see Green and Stokey (1983) and Lazear and Rosen (1981).

8. The information on state pollution prevention laws was obtained from the National Pollution Prevention Roundtable (1996) and through state Web sites and phone calls with state agencies. Further information on the state laws is available from the author upon request.

9. The measure of toxic releases used is "on-site releases." *On-site releases* include all releases to air, water, land, and underground injection. They do not include transfers off-site for recovery or storage.

10. *Waste* differs from *releases* in that it includes quantities of chemicals used for energy recovery, recycling, and treatment. Thus waste includes all chemicals that are generated during the production process and are not embedded in the final product.

11. The programs they evaluate include MBR, but also include other voluntary pollution-prevention programs that range from recognition programs to technical assistance programs. Thus, the fact that they find no evidence that pollution-prevention programs had an effect is not a test of whether MBR had an effect.

12. Helfand et al. (2003) provide an excellent summary of this research, with the exception of research on the effects of instrument choice on rates of technological change; this last group of studies is summarized in Jaffe et al. (2003).

13. Regulations that target inputs include Maximum Attainable Control Technology standards (MACTs) for hazardous air pollutants (U.S. EPA 2000), the technology mandates embedded in the New Source Review (NSR) (U.S. EPA 2002a), or the lead phase-down program for gasoline that targeted the lead content of gasoline (Nichols 1997). Regulations that target emissions include the SO_2 tradable permit system (Stavins 1998) and the RECLAIM air quality program in Los Angeles (Harrison 1999). A recent attempt to regulate based on concentrations is the Water Quality Trading Policy (U.S. EPA 2004b).

14. For a rule of thumb measuring whether industry heterogeneity is sufficient to make market-based instruments preferable to uniform standards, see Newell and Stavins (2003).

15. Specifically, chemicals added to the TRI must be "known to cause or can reasonably be expected to cause in humans—(a) cancer or teratogenic effects or (b) serious or irreversible reproductive dysfunctions, neurological disorders, heritable genetic mutations, or other chronic health effects." Chemicals may also be added to this list if they are "known to cause or can reasonably be anticipated to cause ... a significant adverse effect on the environment ..." (U.S. Code 1986).

16. Keohane (2005) provides an example of the type of empirical analysis that can be done when a policy area is identified that is subject to multiple regulatory instruments. He investigates the adoption of scrubbers at plants subject to tradable permit regulations for sulfur dioxide under the Clean Air Act Amendments of 1990 with plants subject to command-and-control regulations for the same pollutant.

References

Arrow, Kenneth J., and Anthony C. Fisher. 1974. Environmental Preservation, Uncertainty, and Irreversibility. *Quarterly Journal of Economics* 88(2): 312–19.

Baumol, William J., and Wallace E. Oates. 1988. *The Theory of Environmental Policy.* Cambridge, U.K.: Cambridge UP.

Bennear, Lori Snyder, and Cary Coglianese. 2005. The Role of Program Evaluation in Environmental Policy. *Environment* 47(2): 18–39.

Coglianese, Cary, and David Lazer. 2003. Management-Based Regulation: Prescribing Private Management to Achieve Public Goals. *Law and Society Review* 37(4): 691–729.

Florida, Richard, and Derek Davison. 2001. Why Do Firms Adopt Advanced Environmental Practices (And Do They Make a Difference)? In *Regulating from the Inside: Can Environmental Management Systems Achieve Policy Goals?* edited by Cary Coglianese and Jennifer Nash. Washington, DC: Resources for the Future, 89–104.

Gomes, James R. 1994. Halfway to ... Where? A Report on the Implementation of the Massachusetts Toxics Use Reduction Act. Mimeo. Boston, MA: Environmental

League of Massachusetts Education Fund.

Goulder, Lawrence. 2000. Environmental Policy Making in a Second-Best Setting. In *Economics of the Environment*, edited by Robert N. Stavins. New York: W.W. Norton.

Graham, Mary, and Catherine Miller. 2001. Disclosure of Toxic Releases in the United States. *Environment* 43(8): 9–20.

Green, Jerry R., and Nancy L. Stokey. 1983. A Comparison of Tournaments and Contracts. *The Journal of Political Economy* 91(3): 349–64.

Grossman, Sanford J., and Oliver D. Hart. 1983. An Analysis of the Principle-Agent Problem. *Econometrica* 51:7–45.

Hahn, Robert W. 1989. Economic Prescriptions for Environmental Problems: How the Patient Followed the Doctor's Orders. *Journal of Economic Perspectives* 3: 95–114.

Harrison, David, Jr. 1999. Turning Theory into Practice for Emissions Trading in the Los Angeles Air Basin. In *Pollution for Sale: Emission Trading and Joint Implementation*, edited by Steve Sorrell and Jim Skea. London: Edward Elgar, 63–79.

Hart, Oliver D., and Bengt Holmstrom. 1987. The Theory of Contracts. In *Advances in Economic Theory, Fifth World Congress*, edited by T. Bewley. New York: Cambridge UP.

Hassett, Kevin A., and Gilbert E. Metcalf. 1995. Energy Tax Credits and Residential Conservation Investment. Working paper 4020. Cambridge, MA: National Bureau of Economic Research.

Helfand, Gloria E., Peter Berck, and Tim Maull. 2003. The Theory of Pollution Control. In *The Handbook of Environmental Economics*, edited by Karl-Göran Mäler and Jeffrey R. Vincent. Amsterdam: North-Holland/Elsevier Science, 249–303.

Holmstrom, Bengt. 1982. Moral Hazard in Teams. *The Bell Journal of Economics* 13(2): 324–40.

Jaffe, Adam B., Richard G. Newell, and Robert N. Stavins. 2003. Technological Change and the Environment. In *The Handbook of Environmental Economics*, edited by Karl-Goran Mäler and Jeffrey R. Vincent. Amsterdam: North-Holland/Elsevier Science, 461–516.

Jung, Chulho, Kerry Krutilla, and Roy Boyd. 1996. Incentives for Advanced Pollution Abatement Technology at the Industry Level: An Evaluation of Policy Alternatives. *Journal of Environmental Economics and Management* 30: 95–111.

Keenan, Cheryl, Joshua L. Kanner, and Diane Stoner. 1997. Survey Evaluation of the Massachusetts Toxics Use Reduction Program. Methods and Policy Report No. 14. Lowell, MA: Toxics Use Reduction Institute, University of Massachusetts, Lowell.

Keohane, Nathaniel O. 2005. Environmental Policy and the Choice of Abatement Technique: Evidence from Coal-Fired Power Plants. Working paper. New Haven: Yale School of Management.

Kleindorfer, Paul R., Harold Feldman, and Robert A. Lowe. 2000. Accident Epidemiology and the U.S. Chemical Industry: Preliminary Results from RMP*Info. Working paper 00-01-15. Philadelphia, PA: Center for Risk Management and Decision Processes, The Wharton School, University of Pennsylvania.

Laffont, Jean-Jacques, and Jean Tirole. 1993. *A Theory of Incentives in Procurement and*

Regulation. Cambridge, MA: MIT Press.

Lazear, Edward P., and Sherwin Rosen. 1981. Rank-Order Tournaments as Optimum Labor Contracts. *The Journal of Political Economy* 89(5): 811–64.

Maxwell, John W., Thomas P. Lyon, Steven C. Hackett. 2000. Self-Regulation and Social Welfare: The Political Economy of Corporate Environmentalism. *Journal of Law and Economics* XLIII: 583–617.

Milliman, Scott R., and Raymond Prince. 1989. Firm Incentives to Promote Technological Change in Pollution Control. *Journal of Environmental Economics and Management* 17: 247–65.

Natan, Thomas E. Jr., Catherine G. Miller, Bonnie A. Scarborough, and Warren R. Muir. 1996. *Evaluation of the Effectiveness of Pollution Prevention Planning in New Jersey*. Alexandria, VA: Hampshire Research Associates Inc.

National Pollution Prevention Roundtable. 1996. *The Source: The Ultimate Guide to State Pollution Prevention Legislation*. Washington, DC: National Pollution Prevention Roundtable.

Newell, Richard N., and Robert N. Stavins. 2003. Cost Heterogeneity and the Potential Savings from Market-Based Policies. *Journal of Regulatory Economics* 23(1): 43–59.

Nichols, Albert L. 1997. Lead in Gasoline. In *Economic Analyses at EPA*, edited by Richard D. Morgenstern. Washington, DC: Resources for the Future, 49–86.

NRC (National Research Council). 1999. *Arsenic in Drinking Water*. Washington, DC: National Research Council.

Olson, Mancur. 1971. *The Logic of Collective Action*. Cambridge, MA: Harvard UP.

Palmer, Karen, Wallace E. Oates, and Paul R. Portney. 1995. Tightening Environmental Standards: The Benefit-Cost or the No-Cost Paradigm? *Journal of Economic Perspectives* 9(4): 119–32.

Pindyck, Robert. 1991. Irreversibility, Uncertainty, and Investment. *Journal of Economic Literature* 29(3): 1110–52.

Porter, Michael E., and Claas van der Linde. 1995. Toward a New Conception of the Environment-Competitiveness Relationship. *Journal of Economic Perspectives* 9(4): 97–118.

Richards, R. Kenneth. 2000. Framing Environmental Policy Instrument Choice. *Duke Environmental Law and Policy Forum* 10: 221–82.

Snyder, Lori D. 2004. Are Management-Based Regulations Effective?: Evidence from State Pollution Prevention Programs. In *Essays on Facility Level Response to Environmental Regulation*. PhD dissertation. Cambridge, MA: Harvard University.

Stavins, Robert N. 1998. What Have We Learned from the Grand Policy Experiment?: Lesssons from SO_2 Allowance Trading. *Journal of Economic Perspectives* 12(3): 69–88.

———. 2003. Experience with Market-Based Environmental Policy Instruments. In *The Handbook of Environmental Economics*, edited by Karl-Goran Mäler and Jeffrey Vincent. Amsterdam: North-Holland/Elsevier Science, 95–141.

Tenney, Heather M. 2000. A Comparison of Voluntary and Mandatory State Pollution Prevention Program Achievements. Master's Thesis, Civil and Environmental Engineering. Medford, MA: Tufts University.

Tietenburg, Tom H. 1990. Economic Instruments for Environmental Protection. *Oxford Review of Economic Policy* 6(1): 17–33.

U.S. Code. 1986. Toxic Chemical Release Forms. Title 42, Chapter 116, Subchapter II,

Section 11023(d).

U.S. Environmental Protection Agency (EPA). 2000. *Taking Toxics Out of the Air.* Document number EPA-452/K-00-002.

————. 2002a. *New Source Review: Report to the President.* http://www.epa.gov/nsr/documents/nsr_report_to_president.pdf (accessed February 14, 2005).

————. 2002b. *The Toxics Release Inventory (TRI) and Factors to Consider When Using TRI Data.* http://www.epa.gov/tri/2002_tri_brochure.pdf (accessed February 14, 2005).

————. 2004a. Proposed National Emission Standards for Hazardous Air Pollutants; and, in the Alternative, Proposed Standards of Performance for New and Existing Stationary Sources: Electric Utility Steam Generating Units; Proposed Rule. *Federal Register* 69: 4652, January 30.

————. 2004b. *Water Quality Trading Assessment Handbook: Can Water Quality Trading Advance Your Watershed's Goals?* Document number EPA841-B-04-001.

U.S. Food and Drug Administration. 2001. *HACCP: A State-of-the-Art Approach to Food Safety, FDA Backgrounder.* http: www.cfsan.fda.gov/~lrd/bghaccp.html (accessed on February 14, 2005).

U.S. Government Accountability Office (formerly the General Accounting Office). 1998. *Program Evaluation: Agencies Challenged by New Demand for Information on Program Results.* Government Accountability Office Report # GGD-98-53.

Weitzman, Martin L. 1974. Prices vs. Quantities. *Review of Economic Studies* 41: 477–97.

4

The Risk Management Program Rule and Management-Based Regulation

Paul R. Kleindorfer

The deadly release of methyl isocyanate gas in Bhopal, India, in December 1984, followed soon thereafter by a release of aldicarb oxime from a facility in Institute, West Virginia, resulted in great public concern in the United States about the potential danger posed by major chemical accidents. This public concern was translated into law in Section 112(r) of the 1990 Clean Air Act Amendments, which sets forth a series of management-based requirements aimed at preventing and minimizing the consequences associated with accidental chemical releases. These requirements provide the statutory basis for EPA's rule on Risk Management Programs for Chemical Accident Prevention (the RMP rule), a classic form of what Coglianese and Lazer (2003) have called *management-based regulation*.

The RMP rule requires that owners or managers of chemical facilities in the United States that store or use more than specified quantities of toxic or flammable chemicals develop internal procedures for assessing and managing hazards related to chemical releases at their facilities. It further requires that worst-case scenarios associated with such releases be shared with emergency responders and the public, and that facilities periodically file a history of their accidents with EPA. This chapter provides an overview of recent studies that draw on accident history data that firms have submitted to EPA and offers a preliminary assessment of the effectiveness of this form of management-based regulation. The initial data reported under the RMP rule covered the period from about mid-1994 through mid-2000, and also provided details on economic, environmental, and acute health effects resulting from accidents at more than 15,000 U.S. chemical facilities. These data show (1) positive associations between facility hazardousness and both accident frequency and severity; (2) positive associations between accident propensity and debt-equity ratios of parent companies; (3) several interrelated associations between accident propensity and regulatory programs in force; and (4) strong associations between facility hazardousness, facility location decisions, observed accident frequencies, and community demographics.

These results not only provide important baseline data on the accident performance of the U.S. chemical industry; they also provide the basis for an initial assessment of the efficacy of the RMP rule as a form of management-based regulation. In this spirit, I begin this chapter by reviewing results to date and assess their value in providing benchmarks for regulators, communities, and insurers on the safety and underlying risk of chemical facilities. Next, I discuss questions and research issues arising from these studies. I conclude with a discussion of the efficacy of the RMP rule as a form of management-based regulation.

Background on the RMP Rule as Management-Based Regulation

As Bennear theorizes in Chapter 3, management-based regulation may improve the environmental performance of firms in two ways: by mandating risk-reduction planning, and by mandating an exchange of information between firms and regulators. The RMP rule requires both. With some exceptions, EPA requires that all facilities storing on-site more than threshold quantities of any of 77 toxic or 63 flammable substances must prepare and execute a risk management program (RMP). RMPs must contain the following elements:

1. A risk management plan detailing the facility's accident prevention program, emergency response program, and hazard assessment along with administrative information about the facility.
2. A hazard assessment to determine the consequences of a specified worst-case scenario and other accidental release scenarios on public and environmental receptors, as well as a summary of the facility's five-year history of accidental releases.
3. An accidental release prevention program designed to detect, prevent, and minimize accidental releases.
4. An emergency response program designed to protect both human health and the environment in the event of an accidental release.

By December 2000, the cutoff date for the analysis that follows, a total of 15,219 facilities had reported information to EPA under the RMP rule. All facilities covered by the rule were to report any accidental releases of covered chemicals that resulted in deaths, injuries, significant property damage, evacuations, sheltering in place, or environmental damage (see Kleindorfer et al. 2003 for further details).

The findings reported in this chapter come from studies in which my collaborators and I have analyzed the data submitted under the RMP rule using the tools and methods of epidemiology and statistics. My purpose in presenting these findings is to summarize the data in a manner useful to practitioners and policymakers, and furthermore to test specific hypotheses about facility characteristics that might signal the accident propensity of these facilities.

Epidemiology is the study of predictors and causes of illness in humans. Its use in studying industrial accidents—termed *accident epidemiology*—has been proposed by a number of researchers (e.g., Saari 1986; Rosenthal 1997; Kleindorfer et al. 2003). Just as applications of epidemiology in public health frequently use demographic or life-style data of human populations to determine factors that might be associated with the origin and spread of specific illnesses, accident epidemiology can use the demographic and organizational factors of chemical facilities subject to the RMP rule to determine factors that have significant statistical associations with reported accident outcomes. Such analysis yields important findings about the accident propensity of facilities in the U.S. chemical industry, and about the consequences of these accidents over the study period (1994–2000). These results are significant not just because of the relative completeness of the data records, but also because they allow analysis—by specific facilities, sectors, processes, and technologies—of the magnitude of the risks faced by communities and insurers from chemical facilities.

To illuminate a series of plausible, relevant, and testable hypotheses concerning predictors of facility safety, consider the conceptual model of predictors of accident frequency and severity shown in Figure 4-1. The following factors are proposed as potential predictors:

1. The characteristics of the facility itself, including facility location, size, and the type of hazard present; as well as characteristics of the parent company of the facility (capital structure, sales, management systems in place, etc.).
2. The nature of regulations in force that apply to this facility and the nature of enforcement activities associated with these regulations.
3. The socio-demographic characteristics of the host community for the facility, which may represent the level of pressure brought on the facility to operate safely and inform the community of the hazards it faces.

Unobserved managerial factors noted in Figure 4-1—investments in protective activity, preparedness and training, and management intent and commitment—may play an important role in observed outcomes. These managerial factors are fundamental drivers of the accident propensity and preparedness of facilities. These factors are also confounded with these outcomes. For example, the direction of the statistical association between the inherent hazardousness of facilities and accident rates is likely to be confounded

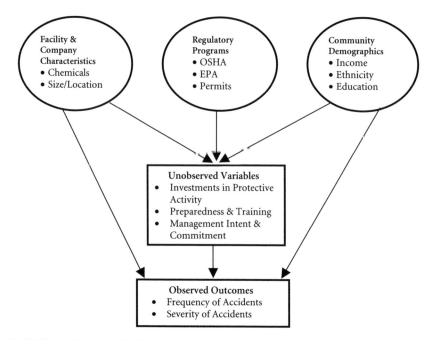

FIGURE 4-1. Framework of Analysis

with unobserved investments in protective activity. On the one hand, more hazardous facilities may give rise to increased accident rates. On the other hand, the greater hazard of these facilities may also stimulate sufficient increases in mitigation investments to counterbalance the impact of the increased hazardousness. The same confounding of the unobserved effects could be elaborated for the other predictors and outcomes in Figure 4-1. Thus, in general, the statistical associations identified in the analysis reported here will reflect the combined effects of regulatory oversight, community pressure, and facility characteristics as these interact with both owner and shareholder governance, and with management systems put in place to monitor and control facility activities.

The central role of management systems, interacting with the factors shown in Figure 4-1, is what makes the RMP rule an interesting and appropriate arena to study the effectiveness of management-based regulation. It is management systems that implement operating procedures for risk management and that structure the organizational processes for balancing mitigation investments and risk reductions. These investments and operating procedures co-determine the accident-related outcomes reflected in Figure 4-1. Such management systems must balance purely economic and financial outcomes with social and environmental outcomes that are more difficult to measure.

As Kagan argues in Chapter 2, the balance referred to here is often stated in terms of the emerging view that profitable companies must ensure both a viable economic franchise as well as a viable social franchise (see also Kleindorfer and Orts 1998; Coglianese and Lazer 2003). A company's economic franchise is viable when sufficient revenues are generated to pay for necessary inputs and production assets, including capital. Its social franchise is viable when communities, regulators, and members of the broader public believe that a company is operating sufficiently within the spirit of the law and supporting social norms to allow it to pursue its economic aims. Either franchise can be threatened by poor environmental performance (as the experience of Union Carbide following the disaster at Bhopal illustrates), as well as poor governance (Enron), poor operational risk control (Barings Bank), and many other functions connected to management and owner oversight and control. In the environmental arena, and especially in environmentally intensive industries such as chemicals, the forces noted in Figure 4-1, singly and jointly, affect both economic and social franchises. The anticipation of these effects should and does demand attention by managers. How this plays out will depend ultimately on the nature of the incentives, positive and negative, arising from the decisions that affect the underlying risks. These incentives are clearly influenced by the nature of the information and management systems put into place in response to management-based regulations.

For example, as noted by many commentators, the requirement that surrounding communities be informed of the risks they face from a facility often generates direct pressure on the facility to reduce those risks (Kleindorfer and Orts 1998; Elliott et al. 2004). A system of management-based regulation effectively reduces the large transaction costs associated with obtaining and legitimating information to back up this pressure. Under such a system, facility management will also face pressure to provide some assurance of information and management system quality. This assurance can be found through third-party auditors, credible others such as insurers, or government inspections. With such a system in place, the dialogue between communities and facilities becomes completely changed. The same may be said about the dialogue between managers and other stakeholders such as employees, insurers, shareholders, and regulators. To the extent that information is backed by credible management systems, as is the requirement under the RMP rule for chemical release hazards, these dialogues become more concrete and legitimate.

How this type of regulation plays out in practice depends upon managers' decision processes. On the one hand, managers face the direct costs of mitigation and monitoring activities; on the other, they can see the benefits of reduced risk of disruptions and accidents to the social and economic franchise of the company. As suggested above, requirements such as the RMP rule, which provide the informational basis and management infrastructure for open communication, may strongly affect how facility managers and owners weigh the costs and benefits of each option.[1] As intuitive as this argument is, however, it must be noted that the costs and benefits of decisions related to environmental, health, and safety policies in a facility or firm are very difficult to measure. An important empirical question is how private sector managers actually respond to the dictates of specific management-based regulations such as the RMP rule. This question motivates the analysis presented in this chapter. The chapter first presents the major findings from data generated by the RMP rule. It then advances observations, based on site visits and focus groups, on the effects of the rule on management systems and community-facility interactions.

Facility Characteristics and Regulatory Impacts

Fifteen thousand, two hundred and nineteen facilities were subject to the RMP rule during the period 1995 to 1999. Managers of these facilities reported a total of 1,945 chemical-release accidents to the RMP database during this period. These accidents resulted in 1,973 worker injuries and $1,018,000,000 in on-site property damage (with business interruption and other indirect costs of such accidents unmeasured, but likely to be much larger than the direct property damages measured). One thousand, one hundred eighty-six facilities (7.8 percent) reported at least one accident (range: 1–15). Of these accidents, 670 (56.5 percent) involved worker injuries (range: 1–69) and 316 (26.6 percent) involved facility property damage (range: $10–$219,000,000). Table 4-1 reports a number of characteristics of facilities, including their location, level of hazard, and regulatory variables of interest. The quantity and nature of chemicals used at each facility are summarized for statistical analyses by a single *total hazard measure*, which is defined roughly as a measure of the hazard of the chemicals on site and the number of covered processes at the facility.[2]

Facility Characteristics and Accident History

What is the association between a facility's characteristics—its geographic region, size, and the chemicals it uses—and its accident history? What is the association between the regulations a facility is subject to and its accidents? I explored these associations with Michael Elliott and Robert Lowe (Elliott et al. 2003). In a separate study with these researchers and Yanlin Wang, I also considered the association between accident propensity and the capital structure of parent companies of facilities as well as various financial measures such as sales (Kleindorfer et al. 2004). I report the results of both these studies here.

The unit of analysis in both of these studies is the facility. We viewed RMP*-Info—EPA dataset containing the RMPs submitted under the RMP rule—as a census of the entire population of relevant facilities. To model the highly skewed outcome data (see Figures 4-2 and 4-3) in a fashion that takes into

TABLE 4-1. Characteristics of RMP Reporting Facilities, Overall and by Whether or Not an Accident Was Reported in RMP (1995–1999)

Geographic region§**	*All Facilities (n = 15,219)*	*No Accident (n = 14,033)*	*1/+ Accidents (n = 1,186)*
% Region I	1.5	1.5	1.7
% Region II	3.2	3.1	3.8
% Region III	5.6	5.4	8.0
% Region IV	15.7	15.5	18.8
% Region V	21.4	21.5	19.5
% Region VI	15.8	15.3	22.0
% Region VII	18.6	19.4	10.3
% Region VIII	6.6	6.8	4.1
% Region IX	8.2	8.3	7.8
% Region X	3.4	3.3	4.2
Number of FTEs**	155	139	345
Number of Chemicals			
Toxic**	1.07	1.04	1.56
Flammable**	.30	.26	.79
All covered chemicals**	1.38	1.30	2.35
Total Hazard**+	13.8	12.7	26.6
% EPCRA-302**	82.2	81.9	85.9
% OSHA PSM**	49.2	46.6	79.2
% CAA Title V**	14.6	13.2	31.9

Notes: * $p < .05$; ** $= p < .01$ by Pearson chi-square or Wilcox on rank test.
+ Methodology for calculating "total hazard" is defined in endnote 2.
§ EPA-defined geographic region. Region I: Maine, New Hampshire, Vermont, Massachusetts, Rhode Island, Connecticut; Region II: New York, New Jersey, Puerto Rico, Virgin Islands; Region III: Pennsylvania, Delaware, District of Columbia, Maryland, West Virginia, Virginia; Region IV: North Carolina, South Carolina, Georgia, Florida, Alabama, Mississippi, Tennessee, Kentucky; Region V: Ohio, Indiana, Michigan, Illinois, Wisconsin, Minnesota; Region VI: Arkansas, Louisiana, Texas, Oklahoma, New Mexico; Region VII: Missouri, Iowa, Nebraska, Kansas; Region VIII: North Dakota, South Dakota, Montana, Wyoming, Utah, Colorado; Region IX: Arizona, Nevada, California, Hawaii, Guam, American Samoa, Trust Territories, Northern Mariana Islands; Region X: Idaho, Oregon, Washington, Alaska.
Source: Elliott et al. (2003).

account not only whether an incident occurred but also the quantity and severity of incidents, we constructed a three-level ordinal scale for the frequency of accidents (0, 1, 2 or more), number of people injured (0, 1, 2 or more), and the amount of property damage (none, $1–$100,000, greater than $100,000). To estimate the relative change in odds of a facility being at level j rather than $j–1$ for a unit change in a covariate, most researchers use extensions of logistic regression models termed proportional odds (PO) models (Agresti 1990). These models assume that this relative change is the same for each change from level $j–1$ to j. When we tested this assumption on these data using a score test, we found that the assumption was generally not valid. Therefore we constructed separate logistic regression models to estimate the odds of being in the lower category versus the upper two and the lower two versus the upper category (Bender and Grouven 1998). We obtained confidence intervals for odds ratios via profile likelihood.

Facilities reporting accidents were more likely to be located in Region III, corresponding to the Mid-Atlantic (8 percent versus 5 percent), Region IV, corresponding to the Southeast (18 percent versus 15 percent), and Region VI, corresponding roughly to the South Central (22 percent versus 15 percent). They were less likely to be located in Regions VII (10 percent versus 19 percent) and VIII (4 percent versus 7 percent), corresponding to the Great Plains and Rocky Mountain regions, respectively. Facilities reporting accidents were also more likely to have more full-time employees (mean = 345 versus 139), to use more toxic and flammable chemicals (mean = 1.6 and 0.8 versus 1.0 and 0.3), and to have higher total hazard measures (mean = 27 versus 13). Facilities with accidents were also somewhat more likely to be regulated under the Right-to-Know Act (EPCRA-302)[3] (86 percent versus 82 percent) and substantially more likely to be regulated under the Occupational Safety and Health Act's Process Safety Management standard (OSHA PSM)[4] (79 percent versus 47 percent) and the Clean Air Act's Title V (CAA Title V)[5] (32 percent versus 13 percent).

Table 4-2 shows, for facilities with at least one reported accident, Spearman correlations between characteristics of each facility and three outcome variables: number of accidents, number of injuries, and property damage. We note that facilities with more full-time employees, more hazardous chemicals in use, and greater total hazard measure were at greater risk of accident, worker injury, and property damage.

Figure 4-2 plots the probability of accident, worker injury, and property damage versus number of full-time employees. The probability of accident increases from less than 3 percent for facilities with fewer than 10 employees to nearly 30 percent for firms with 1,000 employees, then levels off for firms with more than 1,000. The probability of accident actually appears to decline for the very largest facilities (those with 5,000 or more employees), but this decline is not statistically significant. Similar trends are seen for injury risk and property damage risk.

Figure 4-3 plots the probability of accident, worker injury, and property damage versus the total hazard measure for the facility. The probability of any chemical accident during 1995 to 2000 increases from less than 4 percent for firms with a total hazard measure less than five (i.e., the equivalent of five chemicals at twice the threshold level, or one chemical at 32 times—that is, 2^5 times—the threshold level) to approximately 40 percent for firms with a total hazard measure of 50–150. Similarly, the probability of worker injury climbs from about 3–4 percent for firms with a total hazard measure less than 5, then levels off around 30 percent for firms with a total hazard measure of 50–150, and then increases to 50–60 percent as the total hazard measure reaches the 300–400

TABLE 4-2. Spearman Correlations between Number of Accidents, Number of Injuries, and Property Damage in RMP (1995–1999) and Characteristics of Facility, among Facilities ($n = 1,186$) with at Least One Accident

	Number of Accidents	*Number of Injuries*	*Property Damage*
Number of FTEs	.23**	.20**	.14**
Number of Chemicals			
Toxic	.15**	.13**	.04**
Flammable	.10**	.08**	.13**
All covered chemicals	.20**	.17**	.13**
Total Hazard	.13**	.12**	.10**

Notes: * $= p < .05$; ** $= p < .01$.
Source: Elliott et al. (2003).

range. The probability of property damage appears more linearly related to total hazard measure. Results are similar for the more serious outcomes.[6]

In summary, the risk of an accidental chemical release and of attendant worker injuries or property damage increases tenfold as firms grow in size from fewer than 10 to 1,000 full-time employees, then levels off. Similarly, risk of accident, injury, and property damage increases tenfold as "total hazard" measure increases from 0 to 50, then levels off, and then climbs again as total hazard reaches the 300–400 range that characterizes the very largest chemical manufacturers. There are also regional differences among facilities, as well as expected differences in the results of toxics and flammables. Toxic chemicals were more strongly associated with worker injury, whereas flammables were more strongly associated with property damage. This makes sense because fire is obviously capable of causing a much greater degree of damage to property than release of acids or poisonous gases, which are either more contained or less damaging to property although not necessarily less damaging to people.

On the regulatory side, facilities regulated under the Right-to-Know Act had a modestly higher risk of accident, injury, and property damage than other RMP*Info facilities, while facilities regulated under OSHA Process Safety Management (PSM) and CAA Title V had a much higher risk. Nearly all of this excess risk for Right-to-Know and CAA Title V facilities could be explained by their larger size and greater total hazard measures, whereas only about one-half of the excess risk for OSHA PSM facilities could be explained in this manner. Both Right-to-Know Act and CAA Title V target facilities with hazards having significant off-site consequences. The OSHA-PSM standard is focused on on-site hazards, which may not be directly related to inventory levels or numbers of processes, as captured in our hazardousness measure.

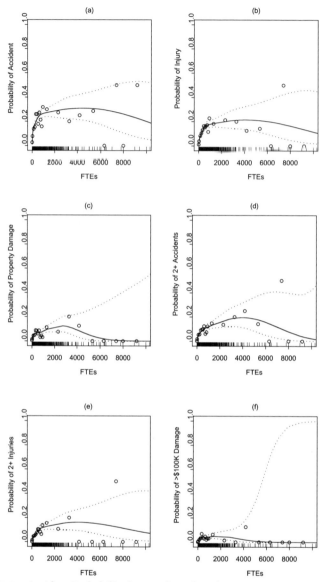

FIGURE 4-2. Accident Probability by Number of Workers

Notes: Probability of having (a) any versus none or (d) two or more versus fewer than two accidents; (b) any versus none or (e) two or more versus fewer than two worker injuries; and (c) any versus no or (f) more than $100,000 versus less than $100,000 in facility property damage in 1995–1999, by number of full-time employee equivalents (FTEs).

Solid line represents mean estimates obtained from cubic spline model with knots at 5, 10, 100, 500, 1,000, and 10,000 employees; dotted line represents associated 95 percent confidence interval. Points are observed percentages for <10, 10–99, 100–199, . . . , 900–999, 1,000–1,999, . . . , 9,000–9,999, and >10,000 employees. Tick marks represent facility FTE measures (truncated at 10,000).

Source: Elliott et al. (2003).

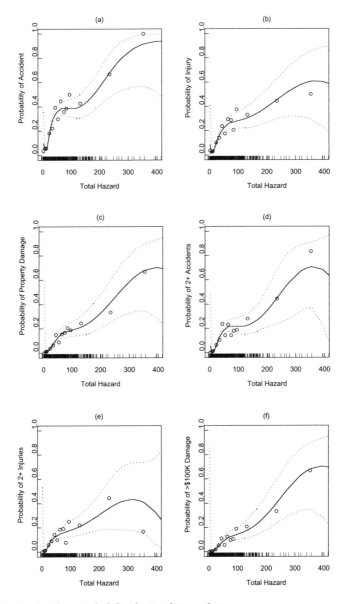

FIGURE 4-3. Accident Probability by Total Hazard Measure

Notes: Probability of having (a) any versus none or (d) two or more versus fewer than two accidents; (b) any versus none or (e) two or more versus fewer than 2 worker injuries; and (c) any versus no or (f) more than $100,000 versus less than $100,000 in facility property damage in 1995-1999, by "total hazard", with total hazard measured as described in endnote 2.

Solid line represents mean estimates obtained from cubic spline model with knots at total hazard measures of 5, 10, 20, 40, and 100; dotted line represents associated 95 percent confidence interval. Points are observed percentages for total hazard measures of <5, 5-10, 10-19,…, 90-99, and >100. Tick marks represent facility total hazard measures (truncated at 400).

Source: Elliott et al. (2003).

Parent Companies and Accident Propensities

We also studied, and report in Kleindorfer et al. (2004), the associations between the capital structure and sales of parent companies and the accident propensities of facilities owned by the parent. In that study we considered four main predictor variables of parent company financial structure and performance: previous year debt/equity ratio, total (net) sales, return on total assets, and return on common equity. We determined *debt-equity ratio* as the ratio of the long-term debt to the common equity. We defined *return on assets* as the ratio of income before extraordinary items divided by total assets minus depreciation and amortization. We defined *return on equity* as income before extraordinary items divided by common equity. To account for the fact that more "intrinsically" hazardous processes tend to involve capital-intensive infrastructure that might confound relationships between attention to safety and financial performance, we used two control variables as proxies for facility hazardousness: the number of full-time-equivalent employees (FTEs), and the "total hazard" measure defined in endnote 2. The key outcome variable in this study was accidents per facility-year in a parent company with specific financial and revenue characteristics, while controlling for the size and hazardousness of the company's facilities.

Concerning predictors related to the financial structure of parent firms (see Kleindorfer et al. 2004), I will just report here the most important and intuitive of the results: parent companies with larger debt-equity ratios tended to be more accident and injury prone, and companies with larger sales tended to be less accident and injury prone. The first result suggests that firms with larger debt-equity ratios tend to behave in a more risk-prone manner relative to safety investments. The second result, fully expected from economic theory, underlines the importance of reputation effects and foregone profits associated with disruptions to operations caused by accidents. In addition, the results suggest that it is not just the current capital structure and sales that are important in determining company and facility approaches to risk management, but that changes in the capital structure and sales in the recent past also have strong associations with accident and injury outcomes. These results imply a number of interactions among RMP management-based regulation, mitigation investments, and risk-taking behavior in firms with different financial structures and different degrees of inherent risk.

Community and Demographic Effects

Environmental justice addresses the issue of whether health risks or environmental impacts from industrial activities are distributed in a manner that comports with basic cultural and social notions of fairness. An extensive body of research in political economy, public policy, and public health has noted associations between environmental and health risks arising from industrial facilities and the socioeconomic status (SES) of host communities (Institute of Medicine 1999). These associations could be caused by firms' anticipating lower levels of collective action and monitoring from communities with lower SES, and therefore preferring to locate hazardous facilities there. These associations could also result from groups of lower SES migrating to sites where such facilities have located, since property values may be lower and thus may be more accessible there. Whatever the reason, a finding that certain communities are at significantly greater risk than others raises fundamental questions for environmental and health authorities.

Empirical findings on the subject of environmental injustice have been mixed. Brown (1995) found that African-Americans and lower-SES Americans are disproportionately likely to live near hazardous waste sites, to be exposed to air pollution or other toxic releases, and not to receive relief from regulatory decision or toxic cleanups. Perlin et al. (2001) found that African-Americans lived closer than whites to the nearest industrial emission source, that African-Americans were more likely than whites to live within two miles of multiple emission sources, and that African-American children five years old and younger were substantially more likely than white children to live near one or more sources of industrial air pollution. Mitchell et al. (1999) found in their examination of South Carolina chemical facilities that, indeed, there are significant negative correlations between the SES of host counties and the risk imposed by chemical facilities, but differences in risk across counties are primarily the result of migration patterns of lower SES individuals to the vicinity of the facilities and not the result of the original location decisions of facility owners.

The debate continues about both the existence and the significance of environmental injustice. But the matter has grown well beyond debate to policy and action. In the United States, terms such as NIMBY (not in my backyard), LULU (locally undesirable land use) and Dumping in Dixie (Bullard 1990) capture the widespread sense that geography and social status have been significant factors in determining which communities end up bearing the environmental risk of manufacturing and waste disposal facilities.

Fueled by public concern and a growing list of documented environmental disasters in the hazardous waste arena, political action by environmental and socially oriented nongovernmental organizations (NGOs) gave rise to national pressure in the United States in the late 1980s to integrate environmental justice considerations into governmental policies. The resulting establishment of the EPA Office of Environmental Justice in 1992 was followed by President Clinton's Executive Order 12898, issued on February 11, 1994, which directed all federal agencies to develop an environmental justice strategy to identify and address adverse health and environmental effects of programs, policies, and activities on minority and low-income populations. These policies have also given rise to fundamental changes in company strategies, extending traditional internal monitoring and assessment procedures to encompass community exposure limits and environmental justice concerns (Moomaw 2001).

Similar activities have been very much in evidence in Europe and Asia, following the disasters in Seveso, Bhopal, and Chernobyl. Citizen activism is also on the rise in emerging economies such as India and China. In the European Union, environmental health monitoring and surveillance systems and regulatory programs have been developed and data to assess the geographic distribution of risk are slowly becoming available. In particular, the Major Accident Reporting System (MARS), set up in 1984 under the Seveso II directive, has the potential to provide data for the European Union that would allow a comparable study to the one reported here (Kirchsteiger 2000).

Using the RMP data together with the 1990 census data, in Kleindorfer et al. (2004) we looked for two essential sources of risk to surrounding populations: (1) risks associated with the decision about where to locate hazardous facilities, which we term *location risk*; and (2) risks associated with the methods of operation and standards of care that are used in existing facilities, which we term *operations risk*.

Our analysis proceeded by first considering the association between community characteristics and location risk—the risk of an intrinsically hazardous facility, as reflected by the quantity of chemicals stored there and their potential for harm, being located in a community. We then considered operations risk—that is, the risk of an accident and the resulting bad outcomes, including injuries and property damage. Two questions can be asked about operations risk: (1) whether the demographics of the communities surrounding facilities are associated with risk of an accident; and (2) whether these community demographics are associated with accident risk *after adjusting for location risk.* This second question addresses the issue of whether facilities in low-SES or African-American communities may exercise less caution in operations, even if they have the same amount of hazardous chemicals on site.

As noted in more detail in Elliott et al. (2004), the relationship between chemical facility risk and the demographics of the surrounding community is complex. As shown in Table 4-3, the RMP data are strongly consistent with a finding that heavily African-American counties in the United States experience both greater location risk and greater operations risk. Greater location risk here means more employees and more hazardous chemicals in use at facilities in these counties. Greater operations risk means that facilities in these counties had greater risks of an accidental chemical release and of having injuries associated with the chemical release. Most importantly, as can be seen in Model 2 of Table 4-3, the operations risk for the most heavily African-American counties persists even after accounting for location risk (i.e., even after controlling for the hazardousness of the facility).

The impact of income and poverty on decisions about where to locate chem-

TABLE 4-3. Operations Risk: Adjusted Relative Risk (RR) for Facility Accidents or Injuries in 1995–2000*

	Model 1	Model 2
1–10% African-American (vs. <1%)	**1.60 (1.33–1.91)**	1.21 (0.99–1.47)
10–20% African-American	**1.79 (1.41–2.29)**	1.19 (0.92–1.54)
> 20% African-American	**3.03 (2.40–3.83)**	**1.85 (1.45–2.37)**
Median income $20K–30K (vs. <$20K)	**1.58 (1.16–2.16)**	0.92 (0.67–1.28)
Median income $30K–40K	**2.05 (1.44–2.94)**	0.99 (0.68–1.44)
Median income $40K+	**2.34 (1.42–3.89)**	1.00 (0.60–1.67)
5–10% income below poverty (vs. <5%)	0.91 (0.64–1.30)	0.80 (0.57–1.13)
10–20% income below poverty	1.01 (0.68–1.49)	0.79 (0.52–1.13)
> 20% income below poverty	0.82 (0.42–1.61)	0.54 (0.28–1.04)
Income Inequality** .4–.45 (vs. <.4)**	1.24 (.88–1.76)	1.21 (.86–1.71)
Income Inequality .45–.55	**1.46 (1.00–2.14)**	1.44 (0.99–2.10)
Income Inequality >.55	**2.08 (1.05–4.24)**	1.84 (0.93–3.65)
10+% Manuf. (vs. <10%)	**1.57 (1.29–1.91)**	**1.30 (1.06–1.59)**
10K–50K Total population (vs. <10K)		**1.61 (1.16–2.26)**
50K+ population		**2.30 (1.64–3.28)**
Number of FTEs (1,000s)		**1.68 (1.44–1.99)**
Total Hazard Measure***		**1.05 (1.05–1.06)**

Notes: * 95 percent confidence intervals in parentheses; bold-face values significant at $p < 0.05$.
** Gini index of income inequality. *** "Total hazard" is calculated as defined in endnote 2.
Source: Elliott et al. (2003).

ical facilities is also complex. Larger facilities are more likely to be located in counties with higher median incomes and higher levels of income inequality, although part of this association is explained by the fact that larger facilities tend to also be located in counties with large populations and large manufacturing labor forces. Similarly, facilities in higher-income counties with higher levels of poverty (that is, high-income-inequality counties) are at greater operational risk as well. However, after adjusting for "total hazardousness," income and income inequality were no longer associated with operations risk.

Thus, higher-risk facilities are more likely to be found in counties with sizeable poor or minority populations that disproportionately bear the collateral environmental, property, and health risks. An alternative, though related, perspective is that communities burdened by low SES or past or present discrimination may be willing to accept these risks in order to obtain the economic benefits of facility location, or that those residents who are not willing to accept this risk move away. For facilities of a similar hazard level, those operated in more heavily African-American counties appear to pose greater risk of accident and injury than those in counties with fewer African-Americans.

Summarizing the Initial Analyses of the RMP Data

Returning to Figure 4-1, we see that the results of the initial analyses of the RMP data are of two types. The first type consists of detailed characterizations of the statistical associations among facility characteristics, parent company characteristics, regulations, and community demographics with accident and injury rates. These associations are for the most part intuitive and support common wisdom in both the risk management community and the financial economics literature. The second type of result to be noted is a corroboration of the overall structure of Figure 4-1. Each of the four factors displayed in Figure 4-1 appears to have strong associations as an independent driver of observed accident and injury rates. This suggests that economic and social franchises both play an important role in company strategy, and that management systems—either voluntarily adopted or triggered by the RMP rule—have responded to a fabric of forces determining the institutional context and incentives faced by company and facility managers. This is a theme I would now like to explore in more detail.

Reflections on the RMP Rule and Management System Effectiveness

Figure 4-4 depicts the general effects expected of management-based regulations such as the RMP rule. The rule itself has requirements for both management systems (the development of a structure to manage and report on the chemical risks involved) as well as for generating and sharing information (in this case, accident history data, worst-case scenarios, and certain other elements of the RMP in the facility). These requirements are expected to result in changes in management systems and, over time through consultants and auditors, should lead to improvements in and more widespread diffusion of best practices. The informational requirements are expected to yield data that reveal characteristics about individual facilities as well as comparative baseline data for specific sectors and technologies. The improved information and standardization of management systems is expected to allow the identification of improvement opportunities as well as to focus the concerns and actions of stakeholders on what is important to them. Finally, these developments in total should lead to observed improvements in outcomes. Expected improvements range from more informed stakeholders, including management and employees of facilities, to improved risk-based pricing of insurance products and directly observable lower accident frequency and severity.

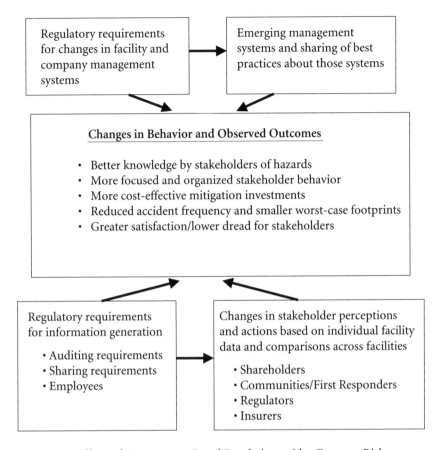

FIGURE 4-4. Effects of Management-Based Regulations with a Focus on Risk Regulations such as the RMP Rule

The general forces depicted in Figure 4-4 have been observed in several other instances. I mention here only the ISO 9000 and ISO 14000 standards, the first for quality management and the second for environmental management systems.[7] Although these are voluntary standards, not regulations, ISO 9000 is a requirement imposed by some governmental agencies on companies competing for public procurement contracts and also imposed by some major manufacturers (such as automotive) on their suppliers. The evolution of these standards is instructive and follows the logic of Figure 4-4.

Consider ISO 9000. In the early 1990s, when the ISO 9000 standard first began to attract attention because of a growing sense that contracting in the new European Union would make the standard a requirement, very little was known about the costs and benefits of ISO 9000. By the mid-1990s, best practices in implementing the standard had been codified, and knowledge about these practices spread quickly around the world. Particular sectors, such as the automotive sector, began to develop more refined versions of the standard for their industries. As standardization proceeded, best practices improved and researchers and practitioners were able to discern what worked and what did not. After about a decade of experience, in many countries it has become clear that there is real payoff, in both financial and operational performance, from an appropriate implementation of ISO 9000 (Corbett et al. 2002). The magnitude of the payoff, including the costs of implementation, depends on the sector, the country of operations, and the experience of the implementing company with process management approaches to quality.

The same dynamic is also occurring with ISO 14000, but because it was just launched in 1996, it is not yet possible to discern the steady-state outcomes from this standard. However, there is a growing expectation that ISO 14000 (and associated auditing standards) will have impacts in the environmental area similar to those achieved by ISO 9000 in the quality arena. Importantly, there are also what I call *synergies of excellence* arising from the joint application of ISO 9000 and ISO 14000. These synergies arise from the fact that both standards are process-based and their effective application requires a deep knowledge of operations at the process level. Accomplishing this pays dividends not just in terms of the focus of the standard (quality or environment), but also in more indirect ways in having employees involved in decision making, in establishing accountability and a culture of management by fact, and in many other ways that have been noted by management scholars as important for operational and financial excellence. These synergies of excellence provide a strong additional, positive externality to improvements in management systems and information. They may also accompany management-based regulations, giving rise to unexpected additional benefits beyond those that were the original focus of the regulation.

Of course, "voluntary" quality and environmental standards may be quite different from management-based risk regulations. Is there any reason to believe that we will see a similar evolution of the above dynamic as a result of recently promulgated management-based regulations, and the RMP rule in particular? I think that there are grounds to believe this, but we are very early in the cycle to have convincing proof. I would adduce in support of the conceptual framework depicted in Figure 4-4 two sets of supporting evidence: first, general trends in the industry related to environmental, health, and safety activities; and second, the results of a number of case studies undertaken by the Wharton Risk Center team that gave rise to the RMP results discussed above.

Concerning general trends, over the past decade there has been an increasing attention to the bottom-line consequences of private sector environmental health and safety (EH&S) activities. This has led, among other things, to a better understanding of the financial consequences of spills and injuries and to innovations in developing management systems that link EH&S activities to both quality, accident, and near-miss management systems and to internal and external financial reporting systems. This work is evident in organizations such as the American Chemistry Council (and its code of Responsible Care), the Chemical Safety and Hazard Investigation Board, and the Center for Chemical Process Safety (CCPS). This increasing attention to consequences has also given rise to significant research efforts designed to understand and codify knowledge about these phenomena, such as those at the Donald Bren School of Environmental Science and Management at the University of California, Santa Barbara; the Mary Kay O'Connor Process Safety Center at Texas A&M University; the Wharton Risk Management and Decision Processes Center at the University of Pennsylvania; and the Department of Environmental Science and Engineering at the University of North Carolina. It is also at the heart of recent initiatives centered around management-based and performance-based environmental regulation (as discussed, e.g., in Coglianese and Lazer 2003). Together, these innovations in practice and the accompanying innovations in research are beginning to point to the necessary ingredients for integrating risk management knowledge, data, and policy objectives in a cost-effective manner. The infrastructure is therefore in place to harvest and document the benefits of management-based regulations such as the RMP rule.

The Wharton Risk Management and Decision Processes Center has documented a number of case studies on the consequences of the RMP rule for both small and large organizations.[8] These case studies demonstrate the nature of changes occurring within chemical facilities, large and small, in response to the rule. First and foremost, the RMP rule is taken very seriously. The image of Bhopal and other major disasters such as those at Longford and Toulouse is very clear to both risk managers and executives; these images remain a driving force in ensuring compliance with the rule. Second, real internal changes have occurred in complying with the rule. These changes include modifications in management systems and a much better understanding of worst-case scenarios and how it is that companies can and must communicate with external stakeholders. Third, the accident history data generated as a part of the RMP rule is understood to be in the "limelight" and considerable care is undertaken to ensure its accuracy. It is understood that these data may be triangulated against first-responder reports, state and federal emergency response centers, and NGOs tracking environmental performance. The fact that the collection and dissemination of these data are taken seriously is important for the quality of the data, but also for the internal changes within operating facilities that are being triggered in response to accidents and incidents that require RMP reports.

Changes from the RMP Process

To indicate the nature of the changes that might result from the RMP process, it is useful to consider the necessary characteristics of management systems in general as contributing factors to process safety. Generally speaking, practitioners believe that in order to reduce the likelihood of low-probability, high-consequence (LP/HC) incidents, it is necessary to have adequate management systems that provide for at least:[9]

1. process designs with the reliability and multiple redundancies appropriate to the consequences of potential process failures;
2. means to identify and address individual LP/HC process safety system component failures (human and material) before they result in HC results;[10]
3. provision of measures to "repair" component failures in a timely and appropriate manner that prevents the unintended introduction of changes that might compromise process safety (management of change guidelines); and
4. measures to ensure that pertinent lessons learned in the facility or in the literature are captured, addressed, and retained in the process management system.[11]

Several organizations—including the CCPS in its Management System Standard, OSHA in its Voluntary Protection Program, and the International Organization for Standardization (ISO)—have attempted to design management system "standards" (by which I mean guidelines, recommended practices, and supporting audit instruments and procedures) that capture the best thinking and experience in promoting and implementing cost-effective approaches to safety and environmental management. An important question arising from these developments is: What are the benefits and limitations of the various management system standards available to guide a company's approach to chemical process safety? Deciding which system to use will require ongoing research to see if facility management systems that follow one or another of these standards exhibit synergies of excellence across relevant performance metrics. The RMP data could provide a stimulus to these developments, by complementing required RMP-based management procedures with added features or with auditing procedures embodied in more detailed management system standards such as the CCPS guidelines, ISO 14000 (in some particular guise), or even ISO 9000 (crafted in a sector-specific fashion to couple quality with risk management). This could facilitate testing hypotheses of the following type for alternative management system standards:

> H1: (Preparedness and Understanding). For facilities with management systems that attempt to be consistent with the requirements of the RMP rule, the level of knowledge of managers and operators about facility hazards and the capabilities for responding to an emergency are directly related to the effectiveness of their process management system as defined by a specific management system standard.
> H2: (Realized Performance). The likelihood and severity of an RMP-defined "process release" at a facility are directly related to the effectiveness of the

facility's process management system as defined by a specific management system standard.

H3: (Worst-Case Footprint). For individual sectors, the scope and magnitude of worst-case scenarios (computed using standardized technologies) at facilities having an RMP-defined "process release" are directly related to the effectiveness of their process management system as defined by a specific management system standard.

The next tranche of data under the RMP rule, collected in 2004, became available for researchers in the summer of 2005.[12] These data will be essential in providing trend analyses to see if RMP is leading to improvements over time, along the lines of those suggested in the hypotheses above. The richer integrated RMP data covering the period 1994–2004 should also be useful in studying whether more specific management system standards (and corresponding audit instruments) may lead to further improvements in preparedness, realized performance, and more accurate worst-case footprints.

Benefits from the RMP Rule

The results reported in this chapter, along with the case studies undertaken as part of the RMP research process by the Wharton Risk Center team, suggest that the RMP rule has already provided some benefits to communities, insurers, and regulators by causing firms to assess, manage, and reveal their environmental and safety risks, and especially to determine and manage factors underlying worst-case scenarios. The data also suggest that the results and the effectiveness of management-based regulation itself are affected by a complicated fabric of facility characteristics, company characteristics, and regulatory and demographic forces (as depicted in Figure 4-1). As noted by Coglianese and Lazer (2003), these interrelationships underline the importance of the public-private balance that is at the heart of integrating the traditional enforcement incentives of regulation with the social and economic interests that co-determine facility and company environmental performance.

Acknowledgments

Research for this paper was partially supported under a Cooperative Agreement with the Chemical Emergency Preparedness and Prevention Office (CEPPO/OEPPR/OEM) of the U.S. Environmental Protection Agency, supporting research at the Wharton Risk Management and Decision Processes Center of the University of Pennsylvania. This chapter summarizes the work of the author and several colleagues, especially Michael Elliott and Robert Lowe, to whom the author expresses a deep debt of gratitude for their contributions to the Accident Epidemiology Project over the past five years. Errors are the sole responsibility of the author. Valuable input on this chapter and additional support was provided by Corporate Associates of the Wharton Risk Center. The author acknowledges helpful comments on earlier drafts from these colleagues and from Howard Kunreuther. Special thanks for comments on and encouragement of this work are due to Jim Belke, James Makris, and Irv Rosenthal.

Notes

1. Besides the work cited in Kleindorfer and Orts (1998) on the effects of specific informational regulations, there are a number of results in the literature on group problem solving supporting the fact that communication and the openness of information on payoffs and constraints confronting participants have strong effects on their actions. See Kleindorfer et al. (1993).

2. More precisely, as introduced in Elliott et al. (2003), the *total hazard* measure used is defined as the sum over all chemicals of $\log_2$ (maximum quantity of inventory on site/threshold), or, alternatively, as the number of chemicals times $\log_2$ of the geometric mean of the maximum-to-threshold quantity ratio. Hence a total hazard measure of 0 indicates that only threshold levels of chemicals are kept in inventory, a measure of 1 means one chemical is kept at up to twice threshold level, 2 means two chemicals kept at up to twice threshold level or one chemical at up to four times threshold level, and so forth; unit changes in this measure can thus be interpreted as either a doubling of volume inventoried of a single chemical or an addition of another twice-threshold chemical on-site.

3. Section 302 of the Emergency Planning and Community Right-to-Know Act lists substances that EPA considers "extremely hazardous."

4. OSHA's standard requires certain companies to adopt a comprehensive management program that addresses technologies, procedures, and management practices.

5. The Clean Air Act Title V requires all large sources of air pollution and many smaller sources of hazardous air pollutants to obtain a federal permit.

6. Adjusting for size of facility reduces but does not eliminate this relationship between total hazard measure and risk of accident, injury, and property damage. Similarly, adjusting for total hazard measure reduces but does not eliminate this relationship between size of facility and risk of accident, injury, and property damage.

7. See Kleindorfer (1997) and Corbett et al. (2002) for a more detailed discussion of these standards.

8. Sanitized versions (to protect the identity of respondents) of these case studies are available from the author on request.

9. I owe these insights to Dr. Isadore (Irv) Rosenthal, former member of the Board of the U.S. Chemical Safety and Hazard Investigation Board (CSB).

10. Many of these concerns are addressed under the rubric of "near-miss management" programs.

11. The CSB has cited various deficiencies in process safety management systems as the "root" cause of most of the accidents it has investigated, and has cited the failure of management systems to incorporate lessons learned into the process management system and make them available to the facility as a major specific contributor.

12. Per information provided by the Office of Emergency Management (OEM), the division of the U.S. EPA responsible for collecting this data. RMP submissions processed to date by OEM suggest that there will be a very significant overlap between reporting facilities for both the earlier and more recent rounds of data collection, allowing trend analysis and many other studies of interest to be conducted using the integrated data from the entire reporting period of 1994–2004.

References

Agresti, Alan. 1990. *Categorical Data Analysis.* New York: Wiley.

Bender, Ralf, and Ulrich Grouven. 1998. Using Binary Logistic Regression Models for Ordinal Data with Non-Proportional Odds. *Journal of Clinical Epidemiology* 51 (10): 809–16.

Brown, Phil. 1995. Race, Class, and Environmental Health: A Review and Systematization of the Literature. *Environmental Research* 69:15–30.

Bullard, Robert D. 1990. *Dumping in Dixie: Race, Class, and Environmental Quality.* Boulder, CO: Westview Press.

Coglianese, Cary, and David Lazer. 2003. Management-Based Regulation: Prescribing Private Management to Achieve Public Goals. *Law & Society Review* 37 (4): 691–730.

Corbett, Charles J., Maria J. Montes, David A. Kirsch, and Maria Jose Alvarez-Gil. 2002. Does ISO 9000 Certification Pay? *ISO Management Systems*, July–August: 31–40.

Elliott, Michael R., Paul R. Kleindorfer, and Robert A. Lowe. 2003. The Role of Hazardousness and Regulatory Practice in the Accidental Release of Chemicals at US Industrial Facilities. *Risk Analysis* 23(5): 883–96.

Elliott, Michael R., Yanlin Wang, Paul R. Kleindorfer, and Robert A. Lowe. 2004. Environmental Justice Revisited: Community Pressure and Frequency and Severity of U.S. Chemical Industry Accidents. *Journal of Epidemiology and Community Health* 58(1): 24–30.

Institute of Medicine. 1999. *Toward Environmental Justice: Research, Education, and Health Policy Needs.* Washington, DC: National Academies Press.

Kirchsteiger, Christian. 2000. Availability of Community Level Information on Industrial Risks in the EU. *Process Safety and Environmental Protection* (*Trans IChemE Part B*), 78(2): 81–90.

Kleindorfer, Paul R. 1997. Market-Based Environmental Audits and Environmental Risks: Implementing ISO 14000. *The Geneva Papers on Risk and Insurance* 22(83): 194–210.

Kleindorfer, Paul R., James C. Belke, Michael R. Elliott, Kiwan Lee, Robert A. Lowe, and Harold I. Feldman. 2003. Accident Epidemiology and the U.S. Chemical

Industry: Accident History and Worst-Case Data from RMP*Info. *Risk Analysis* 23(5): 865–81.

Kleindorfer, Paul R., Michael R. Elliott, Yanlin Wang, and Robert A. Lowe. 2004. Drivers of Accident Preparedness and Safety: Evidence from the RMP Rule. *Journal of Hazardous Materials* 115(August): 9–16.

Kleindorfer, Paul R. and Eric W. Orts. 1998. Informational Regulation of Environmental Risks. *Risk Analysis* 18(2): 155–70.

Kleindorfer, Paul R., Howard C. Kunreuther, and Paul J. Schoemaker. 1993. *Decision Sciences: An Integrative Perspective.* Cambridge: Cambridge UP.

Mitchell, Jerry T., Deborah S. K, Thomas, and Susan L. Cutter. 1999. Dumping in Dixie Revisited: The Evolution of Environmental Injustices in South Carolina. *Social Science Quarterly* 80(2): 229–43.

Moomaw, William R. 2001. Expanding the Concept of Environmental Management Systems to Meet Multiple Social Goals. In *Regulating from the Inside: Can Environmental Management Systems Achieve Policy Goals?*, edited by Cary Coglianese and Jennifer Nash. Washington, DC: Resources for the Future, 126–45.

Perlin, Susan, David Wong, and Ken Sexton. 2001. Residential Proximity to Industrial Sources of Air Pollution: Interrelationships among Race, Poverty, and Age. *Journal of the Air Waste Management Association* 51: 406–421.

Rosenthal, Isadore. 1997. Investigating Organizational Factors Related to the Occurrence and Prevention of Accidental Chemical Releases. In *After the Event: From Accident to Organizational Learning*, edited by Andrew Hale, Bernhard Wilpert, and Matthias Freitag. New York: Pergamon Elsevier Science, 41–62.

Saari, Jorma. 1986. Accident Epidemiology. In *Epidemiology of Occupational Health*, edited by Martti J. Karvonen and M. I. Mikheev. European Series No. 20, World Health Organizations Regional Publications, Copenhagen, 300–20.

5

Environmental Management Under Pressure

How Do Mandates Affect Performance?

Richard N. L. Andrews, Andrew M. Hutson, and Daniel Edwards, Jr.

For more than a decade, corporate leadership groups have advocated a business-led approach to environmental management as an alternative to the rigidity, uneven enforcement, and other imperfections of governmental regulation in a global economy. In preparation for the United Nations Earth Summit in 1992, the World Business Council for Sustainable Development (WBCSD) proposed the development of what became the ISO 14000 suite of voluntary standards for environmental management systems and practices. WBCSD's leaders apparently sought not only to forestall any initiative toward a global environmental regulatory system, but also to "harmonize upward" the basic expectations of good environmental management practices that the most visible firms felt compelled to follow, preventing a "race to the bottom" by competitors in countries lacking the political will or resources to enforce such practices (Schmidheiny 1992). According to these business leaders, sectoral codes of good practice, voluntary certification to international management standards, supply-chain mandates by dominant transnational firms, investor pressures, and simply enlightened corporate self-interest can create more effective pressures than government regulation for environmental and social outcomes that are good both for society and businesses themselves.

These ideas have begun to take root in some firms and sectors. Many firms have voluntarily subscribed to international principles and standards such as the Coalition for Environmentally Responsible Economies (CERES) and Sullivan principles and the ISO 14001 international voluntary standard for environmental management systems. Several trade associations have imposed environmental codes of conduct on their members, such as the American Chemistry Council's Responsible Care© program and the American Forest and Paper Association's required certification of forest management practices. Other sectors have created voluntary environmental certification programs as a market niche, such as "green" hotels. Some investment funds, such as the Calvert Social Investment Fund and the Investor Responsibility Resource Center, have sought to create similar advantages for firms willing to meet specified environmental and social performance criteria.

Dominant suppliers, customers, and vertical integrators in a number of sectors have also begun to impose environmental and other management mandates on their subsidiaries, and on suppliers and customers over whom they have economic influence—so-called *supply chain mandates*. For example, many of the major automotive manufacturers have imposed explicit environmental management system (EMS) requirements on their own facilities and on the facilities of all of their first-tier suppliers, and have encouraged first-tier suppliers to impose similar mandates on their second- and third-tier vendors. Major manufacturers in the electronics sector, and some other sectors, have adopted similar requirements. For example, McDonald's and Kentucky Fried Chicken (KFC) have imposed animal welfare management requirements on their suppliers, and Nike and other firms have imposed labor codes on businesses in their supply chains.

Such mandates provide important opportunities to examine the potential of business-led initiatives as instruments for achieving public purposes such as good environmental management practices (Coglianese 2001). These mandates also raise important questions about the generalizability and possible limitations of such initiatives. Consider requirements that subsidiaries and suppliers of dominant firms introduce a formal—in some cases, third-party-audited—EMS. Such mandates presume two outcomes: first, that manufacturing facilities that adopt EMSs will realize positive environmental benefits and management efficiencies, and second, that these results will occur in facilities that introduce an EMS to comply with a mandate and not only in those that adopt an EMS for their own business reasons. It is conceivable that, as some advocates hope, such private sector mandates will be both more effective and more efficient than government regulations. It is just as plausible, however, that mandates imposed by businesses will come to be seen by those subject to them as being just as burdensome and unproductive as government regulation.

The Business Case: Why Mandate an EMS?

In earlier studies, Andrews and his coauthors (2001; 2003) have provided detailed discussions of the literature on corporate behavior and the natural environment, the business value of pollution prevention initiatives, the relative importance of external pressures versus internal capabilities as motivators (the "resource-based" view of the firm), and the relationships between EMS introduction and changes in environmental performance and regulatory compliance. Here we focus more specifically on issues related to corporate and customer mandates. Why would corporations mandate EMS adoption by their subsidiaries? Why would corporations, particularly dominant firms in supply chains, impose similar mandates on their suppliers? And, ultimately, what effects do such mandates have on the facilities that are subject to them?

Corporate EMS Mandates

Corporate environmental management mandates fall within a far larger domain of issues concerning the conditions under which good environmental management practices are in fact within the self-interest of a business firm. For instance, Richard Andrews and others have argued that "voluntary" environmental initiatives by businesses must logically reflect (1) enlightened self-interest on the part of the firm itself (such as cost-saving pollution prevention initiatives), (2) customer or supplier mandates (such as a mandate to protect brand image or minimize liability, in cases where such business partners hold economic power over the firm in question), or (3) sectoral guidelines and covenants, presumably representing a "club good" that involves reputational risks shared by multiple firms in the sector (R. Andrews 1998; Kollman and Prakash 2002).

At the level of the firm or corporate business unit, the link between environmental management practices and bottom-line business benefits is vigorously debated. Many economists assume that environmental expenditures represent either deadweight costs imposed by government regulations or philanthropy, rather than integral and wealth-generating elements of a core business strategy (Palmer et al. 1995). In contrast, a number of business management scholars have argued that environmental expenditures do contribute positively to market value (Porter and van der Linde 1995a, 1995b; Hart and Ahuja 1996; Klassen and McLaughlin 1996; Sharma and Vredenburg 1998; Dowell et al. 2000; Christmann 2000). Still others have argued that businesses use multiple considerations in defining the bottom line, including, for instance, long-term sustainability of the business as well as short-term profits, strategic change as well as cost minimization for existing products, and both legal and moral responsibilities to stakeholders (employees, suppliers and customers, lenders and insurers, the surrounding community, and others (Evan and Freeman 1988; Freeman 1984). Some even argue that a firm's enlightened self-interest is best served by optimizing a "triple bottom line" rather than a single one by including environmental and social value as well as immediate profits (Elkington 1998; Hart and Milstein 1999; van Heel et al. 2001).

In short, there are conflicting arguments as to what legitimate reasons would motivate a corporation's central management to impose environmental or social management mandates even on itself, let alone on its subsidiary business units, suppliers, or business customers. Some companies, though, are imposing these requirements. Honda, for instance, required all its facilities to implement EMSs by the year 2000, and Canon imposed on all its facilities worldwide not only an EMS mandate but stringent performance requirements such as "zero waste" rules requiring elaborate disassembly of defective products (Honda 2004; Canon 2001). Such mandates impose additional costs that may or may not be productive at the facility level.

What benefits at the corporate level might justify such mandates? Brand and reputation enhancement, minimization of liability, and standardization are three possible reasons. For instance, Canon reportedly faced criticism for pollution at some of its computer chip manufacturing facilities, and used an unusually rigorous overall corporate policy of green responsibility ("zero waste") to isolate and marginalize criticisms of these problems. Honda apparently sought a unified "green" brand image for all its operations as an element of its corporate-level image and marketing. Others have used EMSs to minimize risk of liability from environmental violations or hazards by their subsidiaries, by requiring more standardized and explicit record-keeping and corrective-action procedures.

For motor vehicle manufacturers, and perhaps others, EMS mandates imposed on suppliers also represent elements of broader strategic initiatives to consolidate and optimize complex supply chains. These broader initiatives often include quality-management systems, just-in-time logistics management, integrated electronic tracking and information databases, and the channeling of second- and third-tier supplier relationships through a smaller number of first-tier supplier firms. These initiatives seek to consolidate complex supply chains into more closely monitored and predictable systems. Many corporations may well anticipate business value from receiving consistent information flows from their varied business units and from disseminating best practices from model units to others.

The presence of other pre-existent advanced management practices, particularly quality management, is another influential form of internal capability driving business firms' motivation to adopt an EMS, and perhaps also to impose it on subsidiaries and suppliers (Darnall 2002; Andrews et al. 2003). Several studies have found that nearly twice as many EMS adopters had just-in-time (JIT), total quality management (TQM), or other performance-measurement systems in place than non-adopters (Florida and Davison 2001). In a study of the New Zealand plastics industry, Corbett and Cutler (2000) also reported that preventative approaches and sound internal waste management allowed companies to improve cost control and other management processes.

Some of the pressures for good environmental management practices come from corporate CEOs themselves, who choose to define their own images and reputations, as well as those of their companies, by championing such initiatives. At a personal level, some CEOs genuinely believe in and aspire to "doing well by doing good," and direct their corporations to adopt leadership-level environmental management practices. Some pressure their suppliers and other business partners to do likewise, and some join and lead business leadership groups, such as the World Business Council for Sustainable Development, to promote similar standards throughout their own sectors and in industry more generally. Examples include Chad Holliday at Dupont, Ray Anderson at Interface, and Lord Browne at BP (C. Andrews 1998; Wilson 1998; Andersson and Bateman 2000; Egri and Herman 2000; Holliday 2000). Honda's corporate-wide EMS mandate followed directly from its CEO's initiative; many other corporate environmental mandates have had similar origins. As noted earlier, the ISO 14001 EMS standard itself resulted from an initiative of the WBCSD, whose leaders and members were all corporate CEOs.

Other firms and CEOs, however, have taken the opposite approach. Arguments against corporate environmental mandates include not only short-term profit maximization but also the recognition of differences in environmental conditions, technological constraints, and opportunities for business-related environmental benefits at different facilities and business units, a desire to minimize bureaucratic burdens on subsidiaries (and on suppliers and customers), and the aim of promoting competition and entrepreneurship by each facility to maximize efficiency throughout the supply chain. An important question, accordingly, is why some corporations and CEOs choose to lead in "greenness" and to impose such practices on their subsidiaries, while others do not.

Supply Chain Mandates

Even if it makes sense for businesses to impose corporate mandates on their subsidiaries, a further question is why they should care about the environmental management practices of their suppliers or vendors. What right or interest might justify imposing environmental management mandates on suppliers?

Unlike quality management systems and ISO 9000, environmental management practices concern only the process of production, not necessarily the content or other characteristics of the product itself. Why not just let suppliers compete on product cost and quality, and stand or fall on their own? The same might be asked about labor and human rights mandates, such as those imposed by Nike, Gap, Wal-Mart, and Disney, or the animal-welfare mandates imposed on the farms and livestock operators that supply McDonald's, KFC, and others.

One strong argument in favor of mandates on suppliers in some of these cases (but only some) is that firms in particular sectors face shared threats to their reputations, which in turn creates incentives for them to create sectorwide standards of good practice binding on all their members. Nash (2002) has argued that such trade-association codes have had limited effectiveness over their own members, but appear to be fostering a system of "lead industry regulation" under which dominant firms establish mandates for environmental management practices by the industries and firms that depend on them. That is, sectoral pressures appear more effective in encouraging members to monitor and sanction their suppliers and distributors than pressure from their own peers. Nash notes, however, that more research is needed to learn whether such monitoring and sanctioning programs lead to better environmental results. Lead industry regulation might simply repeat the problems of governmental environmental regulations—problems such as inefficiency, rigidity, and limitations in scope.

A second and equally plausible argument in support of supply-chain mandates is that large, brand-visible firms that market directly to consumers face far greater risk of economic damage to the value of their brand image than do their more anonymous suppliers. They may choose to protect that image through mandates governing labor, human rights, and environmental practices throughout their supply chains. Many such firms have in fact aggressively pursued strategies of maximizing their investment in market dominance through brand-image cultivation while outsourcing the actual production processes, along with their associated labor and environmental liabilities, to separate supplier firms. Ironically, it is the very market power and economic value associated with highly visible global branding that also makes these highly visible, branded corporations most vulnerable to shaming—and, thus, to diminution of the economic value of their brands—if abuses by their suppliers are brought to light. This potential vulnerability is precisely what activist groups and socially conscious pension and investment funds have recently targeted when the brands can be identified with poor practices by the firms that do the actual production (Klein 1999).

On closer examination, then, it appears that firms have the right to set conditions and expectations for their suppliers as well as their subsidiaries, as conditions of doing business with them. They may also have legitimate interests—such as brand value, standardization, or limitation of liability—in doing so. Clearly, to the extent that they represent significant and continuing markets for the supplier firms themselves, large firms also have the power to set conditions and expectations. Environmental management mandates may provide what looks like an easy way for dominant firms to achieve positive image benefits and minimize liability risks. Such mandates may also generate some reductions in production costs, allowing the customer firm to demand commensurate price concessions by the supplier. They may produce some more general management benefits as well, such as increased standardization and rationalization of supply-chain management and perhaps even some environmental innovations (Green et al. 1996; Rondinelli and Berry 1998; Geffen and Rothenberg 2000; Corbett 2002). In a survey of Mexican auto suppliers, Hutson (2001) reported that ISO 14001 was the first formalized management system most Mexican auto suppliers had put in place, and that the vast majority would not have implemented the system without the explicit mandates of the major automakers who were their principal customers (Ford and General Motors). Ninety-six percent of these suppliers also believed they would derive benefits from the implementation of the EMS (Hutson 2001).

An important issue that has only begun to be raised is the extent to which such EMS mandates can or should extend beyond the immediate (first-tier) suppliers to the suppliers' own input vendors, and so on back to the original suppliers of raw materials and energy inputs that go into the supply chain. These earliest stages—mining, logging, smelting, refining, and other raw-material extraction and processing—are often the locus of the most severe environmental impacts in the production chain, but they also are often multiple steps removed from the final product assembly processes that directly supply branded manufacturers. In addition, raw-material commodities often are far more difficult to trace to their producers than are value-added products. On the other hand, if supply-chain mandates are limited to first- and second-tier relationships, they may simply create incentives to outsource and thus externalize environmentally damaging activities one step further in the supply chain, through a new set of real or sham intermediary suppliers. If the social goal is to protect the environment, it should not be subject to manipulation merely by shifting damaging activities to a different locus of ownership in an interdependent chain of business relationships.

Mandates Seen from Below: The Facility's Perspective

So far we have explored reasons why a business might choose to mandate an EMS or other environmental practices or outcomes on the part of its subsidiaries or suppliers. However, it is also worth considering what such mandates mean to the business units affected by them. Private sector mandates, such as those imposed by corporate headquarters or dominant business partners, may look like "regulating from the inside" from a public policymaker's perspective (Coglianese and Nash 2001), but they may look just as much like "regulating from the outside" to the managers of the specific business units to which they apply.

One of the principal arguments in favor of EMSs is that, by establishing more systematic management procedures and explicit objectives for improvement, EMSs will in fact produce benefits to the adopting facility itself. For instance, an EMS should produce improved environmental performance and regulatory compliance, reduced costs and more efficient operations, and more general management benefits to the adopting facility, not just to the corporate parent or customer firm (Andrews et al. 2001). If this argument is correct, then EMS adoption may in fact be in the subsidiary's or supplier's own self-interest, as well as that of the mandating corporation or customer. Empirical studies to date report modest but predominantly positive impacts of EMSs on environmental performance (Berry and Rondinelli 2000; Rondinelli and Vastag 2000; Ammenberg 2001; Melnyk et al. 1999; Hamschmidt 2000; Florida and Davison 2001; Mohammed 2000; Anton et al. 2004; Andrews et al. 2003). Some studies also find improvements in operational efficiencies and other management benefits, or expectations of such benefits, whether or not they improve environmental performance (Chin and Pun 1999; Nash et al. 2000; Steger 2000; Florida and Davison 2001; Hutson 2001; del Brio et al. 2001; Morrow and Rondinelli 2002; Darnall et al. 2002; Andrews et al. 2003). Notwithstanding this literature, there remains relatively little systematic empirical literature on the economic benefits and costs of introducing EMSs to businesses that adopt them.

Of course, introducing an EMS might well bring about little or no benefit, either in environmental performance or in other management improvements. Even if there are some gains, they might be less than the costs of developing and maintaining an EMS. If so, then an EMS mandate will be at odds with the subsidiary or supplier's own business interests, and may appear to such facilities just like another government regulation. The EMS might then merely articulate what a facility was already doing. A facility that is already achieving superior environmental performance might use an EMS simply to document its pre-existing commitments (Coglianese and Nash 2001, *17*), and a low-performing facility might simply implement a "paper EMS"—just to satisfy a corporate or customer mandate according to its letter rather than its spirit—without making any real commitments to significant improvement. For facilities that have only limited environmental impacts, or that have already captured most of the "low-hanging fruit" of remedying economically inefficient environmental practices, or that are small enterprises in which detailed management procedures and documentation add less value than in larger and more complex organizations, the costs of implementing an EMS could be greater than the economic benefits realized.

Finally, whatever the facility's previous level of performance, the effects of an EMS might be expected to vary depending on the organization's motivations and goals in adopting it. A facility that introduces an EMS as a tool to improve the overall efficiency of its use of materials and energy—or to improve its management processes more generally—might well achieve greater improvement in performance outcomes than one that uses it merely to satisfy a corporate or customer mandate in form only, or that wants certification merely to promote its public image (Coglianese 2001).

More generally, the individual production facility, unless it is a free-standing firm in itself, typically lacks full decision authority over many key elements of its environmental management options. The facility may have discretionary authority to change its waste-management procedures, day-to-day operations, and emergency response practices, for instance, but it may not have similar authority to make capital investments in modernizing production processes or modifying products to reduce air pollutant emissions or wastewater discharges, or even to adopt "green accounting" procedures to disaggregate and separately manage the hidden environmental costs of its various products and production lines. Product designs, capital investment decisions, and even accounting practices may themselves be dictated by other corporate mandates. A corporation or customer may mandate that a facility develop an EMS, but if it does not address these more fundamental constraints on the individual facility's authority to adopt greener practices, the "continual improvement" mandated by the EMS may well be limited to relatively marginal refinements.

In short, the interests of the individual business unit almost inevitably conflict at least in part with those of its parent corporation as well as those of its business partners. These conflicts represent a principal-agent problem confronting the corporate headquarters or the customer firm, because in responding to an EMS mandate the applicable business units' competitive opportunities, constraints, and even continued survival may be at stake.

A dominant customer usually has the option of changing suppliers, and a corporation's central management often has the option of making strategic decisions to adopt "greener" product and process innovations, to outsource (and thus externalize) environmentally damaging elements of its supply chain, and even to reconstitute its entire portfolio of business units over time. But the individual production facility typically is far more tightly bound to the sunk costs of particular production technologies, locations, and workforces. Any corporate-level decision on investing capital in modernization of its production processes or products might well lead to the transferring of production to a new factory at another location (China, for instance, in the case of U.S. manufacturing operations) rather than to modernizing an existing facility. It would not be surprising, therefore, if EMS adoption at the facility level were to produce only marginal refinements in existing operations rather than more systematic and strategic greening.

To summarize: to date there have been modest but growing pressures on many businesses, from a variety of stakeholders, to improve environmental management practices—specifically to introduce more formalized environmental management systems. A variety of arguments have been proposed for and against mandating good environmental management practices generally, and EMS adoption specifically. The results of these mandates for environmental and management performance, and their associated costs and benefits, have not yet been empirically well demonstrated.

Study Design and Facility Characteristics

To gain a better understanding of the effects of business-led approaches to environmental management, and particularly the effects of corporate and customer

TABLE 5-1. Response Rates (by sector; $n = 617$)

Sector	Rate (%)
Chemical preparations	19.95
Plastics products	16.42
Metal coatings & allied services	17.82
Motor vehicle parts & accessories	25.00

mandates, we surveyed 3,189 manufacturing facilities in four industrial sectors during the spring of 2003. The survey collected information on environmental management activities and practices, the formal objectives of those activities, changes in environmental performance indicators, and perceptions of associated costs and benefits. The survey instrument also asked about motivations for the adoption of environmental management practices, degree of employee involvement, and facility demographics such as size and ownership status.

Facilities were drawn from four industrial sectors selected for a large number of ISO 14001 registrations, significant presence in the automotive industry (where corporate and supplier mandates are prevalent),[1] and sectors known to have substantial impacts on the natural environment. These included facilities in metal coatings and allied services (SIC3479), chemical preparations (SIC2899), motor vehicle parts and accessories (SIC3714), and plastics products (SIC3089). In each sector, as an indicator of significant environmental impact, we targeted facilities that exceeded reporting thresholds for EPA's Toxics Release Inventory (TRI).

Overall, 617 facilities from these four sectors responded to our survey, yielding an overall response rate of approximately 20 percent (see Table 5-1). Of the four sectors we sampled, motor vehicle parts and accessories had the highest response rate (25 percent) and plastics products the lowest (16 percent). The relatively higher rate of response by the motor vehicle parts sector is not surprising, as recent high-profile efforts by the automotive industry as a whole may make the issue of environmental management activities more salient to managers at these facilities.

The facilities responding to the survey had an overall mean size of 320 employees each, with a median of approximately 150 (Table 5-2). The size of facilities within the motor vehicle parts sector was significantly larger than facility size in the other three sectors, with a median of 340 employees (mean = 607, suggesting a few much larger facilities). Facilities in the other three sectors were not significantly different in employment from one another, although the chemical preparations sector had somewhat fewer employees per facility (with a median of 62 employees, and mean of 145). On average, facilities in all sectors were larger (in terms of mean number of employees) than national averages for those sectors, but they were in line with our expectations for facilities with emissions that exceeded reporting thresholds for TRI.

TABLE 5-2. Facility Size (number of employees, by sector; $n = 617$)

Sector	Mean	Median
Chemical preparations	145.12	62
Plastics products	221.70	125
Metal coatings & allied services	258.21	104
Motor vehicle parts & accessories	607.00	340

TABLE 5-3. Ownership Status (by sector)

Sector	Publicly traded (%)	Privately held (%)	Other* (%)
Chemical preparations ($n = 156$)	44.23	46.79	8.97
Plastics products ($n = 149$)	22.82	71.81	5.37
Metal coatings & allied services ($n = 122$)	37.70	53.28	9.02
Motor vehicle parts & accessories ($n = 172$)	58.14	29.07	12.79

* as reported by respondents

We asked the facilities whether they were privately held or publicly traded, to test the possibility that differences in ownership status might be related to differences in mandating practices and responses (Table 5-3). The greatest disparities in ownership status were between motor vehicle parts and plastics producers: 72 percent of the plastics manufacturers were privately held, whereas 58 percent of the motor vehicle parts manufacturers were publicly traded corporations. Across all four sectors we examined, 49 percent of the facilities in this study were privately held, while approximately 42 percent were publicly traded, and 9 percent asserted some "other" ownership status.

Whether public or private, almost three quarters of all facilities were part of a larger business organization (Table 5-4). Such differences would obviously be relevant to the presence of corporate mandates, and might be associated with differences in responses to customer mandates as well. The greatest difference among sectors on this characteristic was between plastics products, where 60 percent of facilities were part of a larger organization, and motor vehicle parts and accessories, where more than 87 percent reported this organizational status. The high percentage for motor vehicle parts manufacturers was expected, as it reflects larger trends within the industrial organization of the automotive sector as a whole. In this sector, the large automobile manufacturers coordinate complex networks of suppliers that are responsible for the manufacture of complete vehicle subsystems and are themselves often large multinational organizations.

An important variable in this study is the presence or absence of a formalized environmental management system. This variable requires careful construction, however, inasmuch as it involves the presence of multiple elements and cannot be determined merely by asking whether the facility is

TABLE 5-4. Facilities that are Part of a Larger Organization (by sector)

Sector	Part of larger organization (%)
Chemical preparations ($n = 156$)	76.88
Plastics products ($n = 155$)	60.00
Metal coatings and allied services ($n = 121$)	71.01
Motor vehicle parts & accessories ($n = 175$)	87.43

TABLE 5-5. Facilities with a Formal EMS

Overall ($n = 599$)	45.25%
By sector	
Chemical preparations	45.96%
Plastics products	29.22%
Metal coatings and allied services	26.83%
Motor vehicle parts & accessories	73.96%
By ownership status	
Publicly traded	63.37%
Privately held	28.61%
By organizational structure	
Part of larger organization	56.38%
Not part of larger organization	14.94%
ISO 14001-equivalent EMS	26.40%

ISO-certified. Only a small fraction of the facilities that consider themselves to have a formal EMS, and that arguably have all the elements of a formal EMS, have as yet chosen to certify to the ISO 14001 international voluntary standard. By asking each facility not just whether it had a formal EMS, but also to identify all the specific environmental management elements it had in place, we were able to determine the EMS status of 599 of the 617 respondents to our questionnaire.

Table 5-5 shows that of these facilities, 45 percent had what we defined as a formal environmental management system in place, while 26 percent met the more rigorous requirements to be categorized as having an "ISO 14001 equivalent" EMS. The motor vehicle parts and accessories sector had the largest proportion of facilities with a formal EMS in place (74 percent). Again, this large percentage is consistent with our expectations, as both customer and corporate mandates have been introduced throughout the automotive sector during the past several years.

In relating EMS adoption to other corporate characteristics, we found that significant differences were associated with ownership and organizational status. Sixty-three percent of facilities that belonged to publicly traded firms had an EMS in place, while only 29 percent of privately owned facilities did. Fifty-six percent of facilities that were part of a larger organization had a formal

EMS, compared with only 15 percent of facilities that were not part of a larger organization.

For purposes of this study, we considered two broad categories of dependent variables: those estimating environmental performance changes, and those gauging perceptions of the costs and benefits of environmental management practices. We asked managers to indicate whether the facility's performance on 17 environmental indicators, such as energy and water use and hazardous waste generation, had increased, decreased, or not changed over the past three years. We also asked managers to consider the costs of environmental activities of their facilities relative to their benefits, and whether specific types of benefits—such as cost savings and management efficiencies—had been realized from those activities.

For independent variables, we used two measures of the degree of formalization of a facility's EMS, and also asked for information to determine the presence or absence of corporate, customer, or other business-to-business pressures to adopt environmental management systems or practices. By our definition, a facility had a "formal" system if it reported employing a minimum set of specific practices that together could reasonably be called a formal EMS. It had a more rigorous ISO 14001–equivalent EMS if it reported having all practices in place consistent with those required by the ISO 14001 standard (whether or not it was in fact registered to the standard). Facilities whose practices fell short of meeting either of these definitions were classified as not having a formal EMS in place, even if respondents stated that they did. We identified the presence of business-to-business mandates to adopt environmental management systems by asking facilities whether specific external pressures contributed to their decision to adopt them, including separate options for customer and corporate mandates.

Since the dependent variables are categorical, we used chi-square analyses to test for differences in motivations, performance outcomes, and cost-benefit perceptions between facilities. However, in cases where the number of reporting facilities in a particular group or category was not sufficient to meet the assumptions of the standard chi-square test, we used Fisher's exact test. Significance of the chi-square or Fisher's test indicated that a clear difference existed between comparison groups. In using these tests, we are not attempting to attribute causality, but simply to identify clear differences between groups possessing different attributes.

Results

In our first analysis, we compared facilities that reported having in place *both* a business-to-business mandate to adopt an EMS and a formal EMS with those facilities that had *neither*. Do facilities that have a mandate to adopt an EMS, and that actually implement a formal system, report more improvements in environmental performance than those that have neither? And do these facilities perceive these systems as delivering positive benefits, or at least benefits greater than costs, when compared with those that do not have a mandate or a formal EMS in place?

The responses to these questions showed clear and consistent differences in direction of environmental performance change, though only modest differences in magnitude, between these two groups (Table 5-6). Across nearly all indicators, facilities that had mandates and EMSs in place were consistently more likely to report environmental performance improvements than were those that had neither. Most of these differences, however, were modest in their magnitude. For most indicators of environmental performance, the presence of mandates, and even of EMSs, did not appear to be associated with significant differences in the numbers of facilities reporting performance improvement.

What is at least equally interesting, however, is the types of performance indicators for which significant differences *were* evident. These included decreases in energy use, increased use of recycled inputs and recycling of waste, decreases in hazardous and non-hazardous waste generation, and decreased incidence of severe leaks or spills. They did not include other indicators of environmental performance that represented at least equally important and heavily regulated types of environmental impacts in these industrial sectors, such as air and water pollution—not to mention water use, chemical and other material inputs, odors and noise, soil contamination, and even regulatory violations.

The most plausible apparent reason for this discrepancy is that most of the indicators that did show significant differences between the mandate/EMS group and the others represented activities over which managers and employees have direct and discretionary control at the facility level. Energy conservation, recycling, waste management, and leaks and spills are all activities that can be improved incrementally through facility-level management decisions. They may also be easiest and least costly to achieve (representing the "low-hanging fruit"). In contrast, reducing air and water pollution—even though at least equally important environmentally, and often far more so—are likely to require changes in more fundamental product formulations and production processes, or more costly capital investments, all of which are more likely to require corporate-level decisions and commitments.

In contrast to the mixed and mostly modest indicators of environmental performance change, nearly all indicators of perceived benefits of environmental management activities showed significantly more positive outcomes for facilities that had EMSs and mandates than for those that had neither (Table 5-7). These facilities reported highly significant differences in ability to reach new markets, in improved company image, in competitive advantage, and in management efficiencies. They also reported significant differences in cost sav-

TABLE 5-6. Environmental Performance Changes (mandate and EMS vs. neither, past three years)

Environmental activity	B2B mandate/EMS in place n ≈ 192			No B2B mandate/No formal EMS n ≈ 79			sig.
	Increased	Unchanged	Decreased	Increased	Unchanged	Decreased	
Water use	0.22	0.31	0.47	0.23	0.38	0.38	**
Energy use	0.26	0.27	0.47	0.37	0.35	0.27	**
Recycled inputs	0.61	0.30	0.09	0.37	0.52	0.11	**
Recycling of waste	0.72	0.20	0.08	0.45	0.48	0.07	***
Chemical inputs per unit output	0.07	0.38	0.55	0.08	0.47	0.45	
Total material inputs	0.29	0.37	0.34	0.30	0.36	0.34	
Hazardous waste generation	0.09	0.17	0.74	0.12	0.38	0.50	***
Non-hazardous waste generation	0.19	0.24	0.57	0.22	0.45	0.33	***
Wastewater effluent	0.17	0.35	0.48	0.15	0.42	0.43	
Air pollution emissions	0.08	0.34	0.58	0.12	0.43	0.45	
Greenhouse gas emissions	0.08	0.50	0.42	0.07	0.57	0.36	
Noise generation	0.07	0.60	0.33	0.05	0.74	0.21	
Smell generation	0.07	0.53	0.40	0.03	0.67	0.30	
Disruption of the natural landscape	0.04	0.79	0.16	0.10	0.88	0.02	*
Soil contamination	0.02	0.63	0.35	0.03	0.81	0.16	
Severe leaks or spills	0.02	0.39	0.59	0.05	0.61	0.34	**
Violations or potential violations	0.05	0.50	0.45	0.10	0.55	0.35	

* $p \leq 0.05$, ** $p \leq 0.01$, *** $p \leq 0.001$

ings, market share, productivity, and access to capital markets. In fact, the only indicators that did not show significant business benefits for facilities with EMSs and mandates were product differentiation and avoidance of noncompliance penalties (understandably, since the data also showed no significant differences in compliance rates).

In short, there appear to be significant differences in business benefits between facilities subject to a business mandate to adopt a formal EMS that actually do adopt such a system, and those that have neither a formal EMS nor a mandate to do so. These business benefits appear in the survey responses even though relatively few environmental performance changes were reported.

In a further analysis, we compared facilities that had both an EMS and a business-to-business EMS mandate with those that had an EMS but no mandate. The purpose of this comparison was to understand what difference a mandate itself makes. Are there differences in reported performance and perceived benefits between facilities that adopt EMSs in the presence of an outside mandate and those that adopt EMSs voluntarily?

For the most part, our analysis revealed few differences between EMS adopters with and without mandates. None of the 17 environmental performance indicators showed significant differences between these groups. This result suggests that the source of motivation to adopt an EMS—whether it is internal or external to the facility—may not make much difference to how a facility performs environmentally.

Though there were no significant differences between groups with respect to environmental performance indicators, facilities did report several areas where the presence of a mandate appeared to provide benefits (Table 5-8). To a greater degree than the facilities that voluntarily adopted an EMS, facilities with an EMS and a customer mandate reported benefits in the areas of avoided penalties, increased market share, the ability to reach new markets, and product differentiation.

Finally, we compared facilities that did not have the minimum elements of a formal or ISO EMS in place but were subject to a mandate with those that had no EMS and no mandates. We wanted to see if differences in environmental performance or management benefits existed in the presence of customer and corporate mandates even if a formal system had not (or had not yet) been adopted.

The results of this analysis were mixed but interesting. Facilities with a *customer* mandate were significantly more likely to report decreases in accidents and spills than those that were not subject to mandates. Additionally, these facilities more frequently reported constant or decreasing use of material inputs, while those without a customer mandate more frequently reported increases (Table 5-9). Facilities with a *corporate* mandate, on the other hand, much more frequently reported decreases in hazardous waste and regulatory violations than those without such mandates (Table 5-10). One might surmise

TABLE 5-7. Benefits Associated with Environmental Management Activities (mandate and EMS vs. neither)

Benefit Statement	B2B mandate/EMS in place n≈220				No B2B mandate/No formal EMS n≈98				Sig.
	None	Low	Moderate	High	None	Low	Moderate	High	
Cost savings in terms of inputs or taxes	0.48	0.29	0.19	0.04	0.68	0.24	0.08	0.00	**
Avoidance of non-compliance penalties	0.18	0.19	0.31	0.32	0.11	0.16	0.28	0.45	*
Increase in productivity	0.37	0.35	0.24	0.04	0.54	0.24	0.18	0.04	**
Increase in market share	0.51	0.30	0.14	0.05	0.71	0.21	0.07	0.01	**
Ability to reach new markets	0.45	0.28	0.18	0.09	0.72	0.21	0.05	0.01	***
Product differentiation	0.60	0.28	0.07	0.05	0.71	0.20	0.08	0.01	
Improved company/plant image	0.07	0.22	0.46	0.25	0.32	0.26	0.25	0.17	***
Improved access to capital markets	0.59	0.21	0.15	0.05	0.76	0.16	0.07	0.01	*
Improved competitive advantage	0.31	0.32	0.26	0.11	0.67	0.20	0.07	0.06	***
Improved management efficiencies	0.15	0.27	0.45	0.13	0.42	0.31	0.21	0.06	***

$\star\ p \leq 0.05$, $\star\star\ p \leq 0.01$, $\star\star\star p \leq 0.001$

TABLE 5-8. Benefits Associated with Environmental Management Activities (EMS, with and without customer mandate)

Benefit statement	Customer mandate (n ≈ 147)		No customer mandate (n ≈ 48)		Sig.
	None	Some	None	Some	
Avoidance of noncompliance penalties	0.23	0.77	0.04	0.96	***
Increase in market share	0.52	0.48	0.25	0.75	***
Ability to reach new markets	0.60	0.40	0.33	0.67	***
Product differentiation	0.43	0.57	0.20	0.80	*

$^* p \leq 0.05$, $^{**} p \leq 0.01$, $^{***} p \leq 0.001$

from these results that corporate mandates imposed on subsidiaries may tend to emphasize compliance assurance and liability reduction (both likely concerns for a parent firm), whereas customer mandates may tend to focus more on eco-efficiency of production (similarly, a more likely concern for a customer firm). These results are at least suggestive of areas for further investigation and confirmation.

Responses concerning the perceived benefits of environmental management activities revealed several additional areas where facilities with customer mandates differed from those with corporate mandates. Facilities subject to customer mandates were more likely to attribute economic benefits to their environmental management activities, including improved market share and access to new markets, than were those without such a mandate (Table 5-9). These responses may also have reflected the respondents' sensitivity to their customers' expectations in order to maintain access to their existing markets (presumably, that is, the customers imposing the mandate). Facilities with a corporate mandate, in contrast, were no more likely than those without such a mandate to attribute benefits to their environmental management activities.

In short, it appears that there are several types of significant differences in facilities' perceptions of the benefits of their environmental management activities between those subject to mandates and those that were not. Even in the face of such differences, we are reluctant to draw strong conclusions at this point, especially conclusions about causation.

Discussion

Our results suggest a number of possible conclusions about the effects of business-to-business EMS mandates. First, formal environmental management systems appear to be useful tools for achieving modest environmental performance gains, and even more useful for achieving business benefits. Even in

TABLE 5-9. Environmental Performance Changes and Benefits of Environmental Activities (no EMS; customer mandate vs. no customer mandate)

NO EMS

Environmental activity	Customer mandate (n ≈ 167)			No customer mandate (n ≈ 92)			Sig.
	Increased	No change	Decreased	Increased	No change	Decreased	
Total material inputs	0.24	0.47	0.29	0.35	0.32	0.33	**
Severe leaks or spills	0.02	0.39	0.59	0.00	0.46	0.54	*

Benefit statement	Customer mandate (n ≈ 147)		No customer mandate (n ≈ 48)		Sig.
	None	Some	None	Some	
Avoidance of noncompliance penalties	0.16	0.84	0.25	0.75	***
Increase in market share	0.53	0.47	0.67	0.33	***
Ability to reach new markets	0.46	0.54	0.50	0.50	***

* $p \leq 0.05$, ** $p \leq 0.01$, *** $p \leq 0.001$.

TABLE 5-10. Environmental Performance Changes (no EMS, with and without corporate mandate)

Environmental activity	Corporate mandate (n ≈ 167)			No corporate mandate (n ≈ 92)			Sig.
	Increased	No change	Decreased	Increased	No change	Decreased	
Hazardous waste generation	0.15	0.22	0.63	0.14	0.37	0.50	*
Violations or potential violations	0.03	0.53	0.44	0.10	0.58	0.32	*

* $p \leq 0.05$, ** $p \leq 0.01$, *** $p \leq 0.001$.

cases where an EMS was mandated, facilities with an EMS reported environmental performance improvements in significantly greater proportion and claimed greater benefits from their environmental management activities than those that did not.

Second, however, such gains seem to be limited to improvements in waste and spill management and energy efficiency, at least in the four industries studied. The gains generally did not appear to include improvements in environmental indicators associated with more fundamental production processes and product characteristics, such as air and water pollution and chemical and other raw material inputs.

The most likely explanation for these results, we suspect, is that the management of wastes, spills, and energy use are areas in which a facility has greatest discretionary authority to make environmental management improvements and perhaps the most obvious cost- and liability-saving opportunities for doing so. In contrast, changes in air emissions, water pollutant discharges, and inputs of chemicals, other raw materials, and water may be far more fundamentally tied to the facility's basic production and waste treatment technologies and its product design constraints. Changes in these technologies may require more complex approval processes by corporate headquarters, not to mention higher-level strategic and capital expenditure decisions, than the production facility itself has authority or motivation to undertake—if indeed such fundamental technological changes are even possible for the products it makes and the processes it uses to make them.

Third, neither EMSs nor mandates appeared to have any impact on compliance with environmental regulations. This finding is consistent with responses to a related question concerning perceived benefits, where there appeared to be no differences in benefits from reduced penalties for noncompliance. It is also consistent with results from a recent study in the United Kingdom (Dahlström and Skea 2002). Having a formal EMS may streamline the compliance-related elements of environmental management (such as monitoring, record-keeping, and inspections), but it may not result in fewer violations.

Fourth, our responses show far more consistent perceptions of business benefits from formalized environmental management activities than perceptions of improved environmental performance. Respondents perceived these benefits principally in terms of business image, market access, cost reduction, and increased efficiency. If correct, this finding appears to confirm some of the potential benefits of EMSs (and of other management system refinements) hypothesized for corporations and customer firms as well as for supplier facilities themselves.

It is also true, however, that in our analyses so far we are dealing with the respondents' perceptions of benefits, and not as yet with concrete measures of them. Respondents with EMSs in place may have become more aware of the potential benefits associated with environmental activities due to the implementation process, and more invested in that process, and may therefore be more likely to list the benefits as such. For respondents facing customer mandates, moreover, the nature of some of the benefits—image, access to markets, and competitive advantage, for example—is implicit in the customer mandates themselves: some benefits may be seen as benefits simply because the customers requiring the systems have made them so.

Fifth, business-to-business mandates by themselves did not seem to produce many significant differences in environmental performance outcomes in facilities that had these mandates compared with facilities that had an EMS but no mandate. This is a hopeful result in the sense that mandates did not appear to distort the value of an EMS—for instance, by creating incentives for "paper EMSs" just to satisfy customer requirements. If the mandates also provide an initial motivation for facilities to undertake an EMS, which in turn provides some real environmental performance and management benefits, that too may be all to the good. At the same time, this finding suggests that public policymakers should perhaps be cautious in relying on business-to-business mandates as instruments of—let alone substitutes for—public policy mandates to protect and sustain the natural environment.

Finally, we found it intriguing that corporate mandates appeared to be more closely associated with performance improvements that reduced liability, while customer EMSs appeared more closely associated with those that reduced costs and increased efficiency (notwithstanding some overlap, particularly on cost savings). Facilities with a customer mandate also reported significantly greater cost savings and greater benefits from improved access to new markets and improved plant image, while cost savings was the only benefit that facilities with a corporate mandate rated higher than facilities without such a mandate. These apparent differences may be worth further investigation.

Implications: Business-to-Business Mandates Generally

Overall, this research suggests a number of plausible insights into how facility managers view EMSs and corporate and customer mandates, and into their strengths and limitations as a strategy for improving environmental quality. Although further empirical research would be worthwhile, the results suggest that at the level of production facilities, where environmental performance impacts are most directly felt, the opportunities for serious "continual improvement" in environmental performance may be far more narrowly constrained by sunk costs in particular production technologies and product characteristics—"technological path dependence"—than idealistic visions of the leverage of voluntary EMS initiatives would suggest. Continual improvement in waste management, incident reduction, and energy conservation may be limited and may not evolve into more fundamental ecological modernization of product design and production processes to reduce air and water pollution and other important environmental impacts.

Making these more fundamental production processes more environmentally benign may require strategic corporate-level adaptations to more sustainable products and production processes over time. Examples of these adaptations include decisions about modernizing or replacing entire production facilities or shifting to more ecologically sustainable suppliers, rather than simply imposing EMS mandates on existing ones. Stuart Hart, for instance, has argued that "continual improvement" procedures such as ISO 14001 will never achieve more than incremental change, and that major environmental performance improvements will result only from the Schumpeterian processes of disruptive innovation and creative destruction across technologies, facilities, and even firms (Hart 1997; Hart and Milstein 1999). Business-to-business drivers of these larger environmental performance improvements most likely will involve not simply EMS mandates imposed on already-existing subsidiaries and suppliers, but also larger patterns of business behavior and the dominant rules, institutions, and incentives of business-to-business decisionmaking, such as finance, investment, insurance, disclosure and due diligence, and mergers and acquisitions (Andrews 2003).

In short, significant ecological modernization of most industrial production processes may require powerful and persistent market or regulatory forces to change fundamentals. Drivers such as large environment-related increases in input factor prices, strong and persistent forces of consumer and investor demand, or strong and effectively enforced regulatory constraints might be needed to prompt substantial change in business practices. In the absence of strong market forces favoring ecological modernization, this conclusion appears consistent with Robert Kagan's: that large improvements in environmental performance, particularly in established industries, may require periodic tightening of regulatory mandates on all firms in a given sector or market rather than reliance on voluntary initiatives—or even on business-to-business mandates (Kagan 2006).

From a public policy perspective, achieving significant environmental performance improvement may also ultimately require the creation of more effective governance mechanisms for the most environmentally damaging inputs and impacts themselves, at scales ranging from local ecosystems to global trade, rather than simply the promotion of management systems for continual improvement at the margins of normal production and operations in existing facilities. EMS mandates should not be expected to correct environmental impacts that result from externalities or common-pool resource conditions, in which markets and the interests of individual businesses "fail" by definition. Governmental intervention, or some other governance mechanism beyond individual firms, will be necessary to manage such impacts. EMS mandates cannot by themselves correct these market failures, nor can they correct the wide range of environmental impacts that result not from businesses' decisions but from those of governments, individuals, and other nonbusiness organizations.

Notes

1. For example, in 1999, Ford Motor Co. required all of its suppliers to certify at least one manufacturing site to ISO 14001 by the end of 2001 and all manufacturing sites shipping products to Ford by July 1, 2003 (Ford Motor Company 1999). Similarly, General Motors Corporation required all of its suppliers to certify their conformance to ISO 14001 by the end of 2002 (General Motors 1999).

References

Ammenberg, Jonas. 2001. *How Do Standardized Environmental Management Systems Affect Environmental Performance and Business?* Licentiate Thesis No. 907. Sweden: Linköping University, Department of Physics and Measurement Technology.

Andersson, L., and T. Bateman. 2000. Individual Environmental Initiative: Championing Natural Environmental Issues in U.S. Business Organizations. *Academy of Management Journal* 43 (4): 548–70.

Andrews, Clinton J. 1998. Environmental Business Strategy: Corporate Leaders' Perceptions. *Society & Natural Resources* 11: 531–40.

Andrews, Richard N. L. 1998. Environmental Regulation and Business "Self-Regulation." *Policy Sciences* 31(3): 177–97.

Andrews, Richard N. L. 2003. Sustainable Enterprise: Implications for International Finance and Investment. Paper presented at the New America Foundation, February 28, 2003. Washington, DC: New America Foundation.

Andrews, Richard N. L., Nicole Darnall, Deborah Rigling Gallagher, Suellen Terrill Keiner, Eric Feldman, Matthew Mitchell, Deborah Amaral, and Jessica Jacoby. 2001. Environmental Management Systems: History, Theory, and Implementation Research. Chapter 2 in *Regulating from the Inside: Can Environmental Management Systems Achieve Policy Goals?*, edited by Cary Coglianese and Jennifer Nash. Washington, DC: Resources for the Future, 31–60.

Andrews, Richard N. L., Deborah Amaral, Nicole Darnall, Deborah Rigling Gallagher, Daniel Edwards Jr., and Chiara D'Amore. 2003. *Do EMSs Improve Performance?* Final Report of the National Database on Environmental Management Systems. Chapel Hill, NC: Department of Public Policy, University of North Carolina at Chapel Hill. http://ndems.cas.unc.edu/

Anton, Wilma Rose Q., George Deltas, and Madhu Khanna. 2004. Incentives for Environmental Self-Regulation and Implications for Environmental Performance. *Journal of Environmental Economics and Management* 48: 632–54.

Berry, Michael A., and Dennis A. Rondinelli. 2000. Environmental Management in the Pharmaceutical Industry: Integrating Corporate Responsibility and Business

Strategy. *Environmental Quality Management* 9(3) :21-33.

Canon Corporation. 2001. *Environmental Report 2001.* Online at http://www.canon.com/environment/report/report2001e.pdf (accessed December 19, 2004).

Chin, Kwai-Sang, and Kit-Fai Pun. 1999. Factors Influencing ISO 14000 Implementation in Printed Circuit Board Manufacturing Industry in Hong Kong. *Journal of Environmental Planning and Management* 42(1): 123–35.

Christmann, Petra. 2000. Effects of "Best Practices" of Environmental Management on Cost Competitiveness: The Role of Complementary Assets. *Academy of Management Journal* 43(4): 663–880.

Coglianese, Cary. 2001. Policies to Promote Systematic Environmental Management. In *Regulating From the Inside: Can Environmental Management Systems Achieve Policy Goals?*, edited by Cary Coglianese and Jennifer Nash. Washington, DC: Resources for the Future, 181–97.

Coglianese, Cary, and Jennifer Nash. 2001. *Regulating from the Inside: Can Environmental Management Systems Achieve Policy Goals?* Washington, DC: Resources for the Future.

Corbett, Charles J. 2002 (unpublished paper). Diffusion of ISO 9000 and ISO 14000 Certification through Global Supply Chains. Los Angeles, CA: UCLA, Anderson School, 33 pp.

Corbett, Lawrence M., and Denise J. Cutler. 2000. Environmental Management Systems in the New Zealand Plastics Industry. *International Journal of Operations and Production Management* 20(2): 204–32.

Dahlström, Kristina, and Jim Skea. 2002. *Environmental Management Systems and Operator Performance at Sites Regulated under Integrated Pollution Control.* Environmental Agency R&D Technical Report P6-017/2/TR, prepared by the Policy Studies Institute. Bristol, England: United Kingdom Environmental Agency.

Darnall, Nicole. 2002. *Why Firms Signal "Green": Environmental Management System Certification in the United States.* PhD dissertation, University of North Carolina at Chapel Hill.

Darnall, Nicole, Deborah Rigling Gallagher, and Richard N. L. Andrews. 2002. ISO 14001: Greening Management Systems. In *Greener Manufacturing and Operations*, edited by Joseph Sarkis. Sheffield, England: Greenleaf Publishing, 178–90.

del Brio, Jesus Angel, Estaban Fernandez, Beatriz Junquera, and Camilio Jose Vazquez. 2001. Motivations for Adopting ISO 14001 Standards: A Study of Spanish Industrial Companies. *Environmental Quality Management* 10(4): 13–26.

Dowell, Glen, Stuart Hart, and Bernard Yeung. 2000. Do Corporate Global Environmental Standards Create or Destroy Market Value? *Management Science* 46(8): 1059–74.

Egri, Carolyn, and Susan Herman. 2000. Leadership in the North American Environmental Sector: Values, Leadership Styles, and Contexts of Environmental Leaders and Their Organizations. *Academy of Management Journal* 43(4): 561–604.

Elkington, John. 1998. *Cannibals with Forks: The Triple Bottom Line of 21st Century Business.* Oxford: Capstone.

Evan, William M., and R. Edward Freeman. 1988. Stakeholder Theory of the Modern Corporation. In *Ethical Theory and Business*, edited by Tom L. Beauchamp and Norman E. Bowie, 3rd edition. Upper Saddle River, NJ: Prentice Hall. Reprinted in

Ethical Issues in Business: A Philosophical Approach, edited by Thomas Donaldson and Patricia H. Werhane. Upper Saddle River, NJ: Prentice Hall, 1999, 247–57.

Florida, Richard, and Derek Davison. 2001. Why Do Firms Adopt Environmental Practices (And Do they Make a Difference)? In *Regulating From the Inside: Can Environmental Management Systems Achieve Policy Goals?,* edited by Cary Coglianese and Jennifer Nash. Washington, DC: Resources for the Future, 82–104.

Ford Motor Company. 1999. Ford Becomes First U.S. Automaker to Require Suppliers to Achieve ISO 14001 Certification. Press release, September 21, 1999. Dearborn, MI: PRNewswire.

Freeman, R. Edward. 1984. *Strategic Management: A Stakeholder Approach.* Boston: Pitman.

Geffen, C., and S. Rothenberg. 2000. Suppliers and Environmental Innovation. *International Journal of Operations and Production Management* 20(2): 166–86.

General Motors Corporation. 1999. General Motors Sets New Level of Environmental Performance for Suppliers. Press release, September 21, 1999. Detroit, MI: GM Communications.

Green, Ken, Barbara Morton, and Steve New. 1996. Purchasing and Environmental Management: Interactions, Policies and Opportunities. *Business Strategy and the Environment* 5: 188–97.

Hamschmidt, J. 2000. Economic and Ecological Impacts of Environmental Management Systems in Companies: Experiences from Switzerland. EURO Environment, Aalborg, Denmark *(cited in Ammenberg 2001).*

Hart, Stuart L. 1997. Beyond Greening: Strategies for a Sustainable World. *Harvard Business Review* 75(1): 66–76 (Jan.–Feb. 1997).

Hart, Stuart L., and Gautam Ahuja. 1996. Does It Pay to be Green? *Business Strategy and the Environment* 5: 30–7.

Hart, Stuart L., and Mark B. Milstein. 1999. Global Sustainability and the Creative Destruction of Industries. *Sloan Management Review* 41(1): 23–33 (Fall 1999).

Holliday, Chad. 2000. Business Growth and Sustainability – Challenges for the New Century. Nikkei Ecology Magazine Anniversary Symposium of the Global Roundtable on Environment and Business, Tokyo, Japan, May 23, 2000.

Honda Corporation. 2004. *Honda Environmental Annual Report 2004.* Online at http://world.honda.com/environment/2004report/pdf/2004E_report_full.pdf

Hutson, A. 2001. *ISO 14001 and the Automobile Industry in Mexico.* Master's thesis. Durham, NC: Duke University, Nicholas School of the Environment and Earth Sciences.

Kagan, Robert A. 2006. Environmental Management Style and Corporate Environmental Performance. In *Leveraging the Private Sector: Management-Based Strategies for Improving Environmental Performance,* edited by Cary Coglianese and Jennifer Nash. Washington, DC: Resources for the Future, Chapter 6.

Klassen, Robert D., and Curtis P. McLaughlin. 1996. The Impact of Environmental Management on Firm Performance. *Management Science* 72 (3): 2–7.

Klein, Naomi. 1999. *No Logo.* New York: Picador.

Kollman, Kelly, and Aseem Prakash. 2002. EMS-Based Environmental Regimes as Club Goods: Examining Variations in Firms-level Adoption of ISO 14001 and EMAS in U.K., U.S., and Germany. *Policy Sciences* 35: 43–67.

Melnyk, Steven A., Roger Calatone, Rob Handfield, R. L. Tummala, Gyula Vastag,

Timothy Hinds, Robert Sroufe, Frank Montabon, and Sime Curkovic. 1999. *ISO 14001: Assessing Its Impact on Corporate Effectiveness and Efficiency.* Tempe, AZ: Center for Advanced Purchasing Studies, Arizona State University.

Mohammed, M. 2000. The ISO 14001 EMS Implementation Process and Its Implications: A Case Study of Central Japan. *Environmental Management* 25 (2): 177–88.

Morrow, David, and Dennis A. Rondinelli. 2002. Adopting Environmental Management Systems: Motivations and Results of ISO 14001 and EMAS Certification. *European Management Journal* 20 (2): 159–71.

Nash, Jennifer. 2002. Industry Codes of Practice: Emergence and Evolution. In *New Tools for Environmental Protection: Education, Information, and Voluntary Measures,* edited by Thomas Dietz and Paul C. Stern. Washington, DC: National Academy Press, 235–52.

Nash, Jennifer, John Ehrenfeld, Jeffrey MacDonagh-Dumler, and Pascal Thorens. 2000. ISO 140001 and EPA's Region I's StarTrack Program: Assessing their Potential as Tools in Environmental Protection. In *Learning from Innovations in Environmental Protection,* edited by DeWitt John and Rick Minard. Washington, DC: National Academy of Public Administration.

Palmer, Karen, Wallace Oates, and Paul Portney. 1995. Tightening Environmental Standards: The Benefit-Cost or the No-Cost Paradigm? *The Journal of Economic Perspectives* (4): 119–32.

Porter, Michael E., and Claas van der Linde. 1995a. Green and Competitive: Ending the Stalemate. *Harvard Business Review* 73(5): 120–34.

———. 1995b. Toward a New Conception of the Environment-Competitiveness Relationship. *The Journal of Economic Perspectives* 9(4): 97–118.

Rondinelli, Dennis A., and Michael A. Berry. 1998. Strategic and Environmental Management in the Corporate Value Chain at Shaw Industries. *National Productivity Review* (Summer 1998): 17–25.

Rondinelli, Dennis A., and Gyula Vastag. 2000. Global Corporate Environmental Management Practices at Alcoa. *Corporate Environmental Strategy* 7(1): 288–97.

Schmidheiny, Stephen. 1992. *Changing Course: A Global Business Perspective on Development and the Environment.* Cambridge, MA: MIT Press.

Sharma, Sanjay, and H. Vredenburg. 1998. Proactive Corporate Environmental Strategy and the Development of Competitively Valuable Organizational Capabilities. *Strategic Management Journal* 19(8): 729–53.

Steger, Ulrich. 2000. Environmental Management Systems: Empirical Evidence and Further Perspectives. *European Management Journal* 18(1): 23–37.

van Heel, Oliver Dudok, John Elkington, Shirley Fennell, and Franceska van Dijk. 2001. *Buried Treasure: Uncovering the Business Case for Corporate Sustainability.* London: SustainAbility.

Wilson, G. 1998. Sustaining Environmental Management Success: Best Business Practices from Industry Leaders. New York: John Wiley and Son.

6

Mandating Insurance and Using Private Inspections to Improve Environmental Management

Howard Kunreuther, Shelley H. Metzenbaum, and Peter Schmeidler

Under what circumstances can a system of mandatory insurance coupled with private inspections improve a firm's current management of safety and environmental risks? Requiring firms to purchase insurance as a condition for operation holds particular promise as a complement to many government regulatory programs, especially when government has a low capacity for monitoring compliance. By coupling mandatory insurance with well-enforced standards or regulations, government may also be better able to identify misalignments between regulations and risk-reducing management practices.

Section 112(r) of the U.S. Clean Air Act Amendments (CAAA) of 1990 created two new federal regulatory programs aimed at preventing releases of hazardous chemicals: the Process Safety Management (PSM) standard of the Occupational Safety and Health Administration (OSHA) and the Risk Management Program of the Environmental Protection Agency (EPA). These two programs, both of which regulate using management-based approaches (Coglianese and Lazer 2003), face a problem common to many regulatory programs: an inadequate inspection presence.

The OSHA program requires facilities containing large quantities of highly hazardous chemicals to implement accident prevention and emergency response measures to protect workers. Firms are aware that, because of limited personnel, OSHA normally visits only those firms known to have serious workplace problems. In contrast, firms not in this category are estimated to have less than a 1/80 chance per year of being inspected by OSHA (Soltan 2003).

One empirical study of a typical small-business manufacturing sector supports this estimate and illustrates the nature of the problem (Weil 1996). Over a 20-year period, 5 percent of custom woodworking establishments were inspected each year. Sixty percent of the facilities, however, were not inspected even once over the two-decade period. Unless fines for violating a given regulation are exceptionally high, firms have little economic incentive to worry about OSHA checking to see whether they have met PSM standards or other workplace regulations.

The EPA's Risk Management Program requires the same facilities to perform a hazard assessment, estimate consequences from accidents, and submit a summary report to EPA called the Risk Management Plan (RMP). Under the Clean Air Act, fines for noncompliance can reach $27,500 per day. Discussions with EPA personnel, however, suggest that these penalties are rarely imposed by the agency, partly because it does so little monitoring. For example, EPA's Region III office has only five auditors for inspecting its many industrial facilities in Delaware, Maryland, Pennsylvania, Virginia, West Virginia, and the District of Columbia (Kunreuther et al. 2002).

Lower levels of inspection or other effective methods for compliance evaluation tend, not surprisingly, to lead to lower compliance rates (Weil 1996). They also reduce opportunities for government to find and require firms to correct the sorts of risky practices that regulations seek to diminish. Increased inspections have the potential of encouraging firms to adopt safer practices by uncovering areas of potential improvement. High government penalties or cost savings associated with investments in environmental and safety management may counterbalance low inspection rates, but only if firms are aware of the high potential penalties or cost savings and, in the case of the penalties, only if firms believe they will be applied when appropriate.

Reformers have advocated certified third-party inspection programs as a mechanism to deal with low inspection levels, as either a complement or an alternative to traditional regulatory programs (National Academy of Public Administration 2001). ISO 14000 addresses the environmental performance of firms and encourages firms to "minimize harmful effects on the environment caused by [their] activities, and to achieve continual improvement of [their] environmental performance" (ISO 2004). However, there are no specific ISO requirements for firms to adopt specific measures, achieve specific outcomes, or report any of this information to the public. This lack of specificity has led to skepticism by the concerned public about such programs achieving their objectives. Similar concerns have been raised about other types of voluntary environmental management systems (MSWG 2004).

The combination of private inspection and insurance is a potentially powerful one for meeting—often even exceeding—environmental and safety regulations. Suppose an inspection reveals management practices that a company can adopt to reduce risks and hence lower insurance premiums. The company will want to adopt these measures if the expected discounted cost savings exceed the costs of new practices.

Insurers have an economic incentive to conduct inspections that focus on risk reduction because they want to reduce both the likelihood of paying a claim and the size of their payments. The insurer's economic survival depends on estimating the risk of future losses accurately, not on ensuring compliance with government laws. To the extent that conventional regulations are well aligned with risk-reducing behaviors, insurers are likely to uncover noncompliance problems and encourage their correction. To the extent that they are not, insurers have little reason to inspect for regulatory compliance.

In the next section of this chapter, we detail the conditions for insurability of a risk and identify circumstances when voluntary insurance coverage may not be a viable option for managing environmental and safety risks from the insurer's perspective. Next, we examine why prospective buyers may not always make rational decisions about insurance purchases and may even overlook opportunities for cost-effective investments in risk reduction. We then consider the rationale for coupling mandatory insurance with regulations in order to reduce losses associated with safety and environmental risks, and explore the benefits of pairing inspections with mandatory insurance. We offer a possible pilot study with the workers' compensation system in Louisiana to explore and understand how variations in program implementation decisions might affect the value of the program. We conclude by suggesting future issues for clarification and areas for research.

Supply of Insurance and Conditions for Insurability

If insurance is to be useful in complementing regulation, the risks to be addressed must be insurable.[1] What does it mean to say that a particular risk is insurable? This question must be addressed from the vantage point of the potential supplier of insurance who offers coverage against a specific risk at a stated premium. Three conditions must be met before insurance providers are willing to offer coverage against an uncertain event. Condition 1 is the ability, when providing different levels of coverage, to identify and quantify or estimate the chances of the event occurring and the extent of the losses likely to be incurred. Condition 2 is the ability to set premiums for each potential customer or class of customers. This requires some knowledge of the specific customer's risk in relation to that of others in the population of potential policyholders. Condition 3 requires that the premium charged by the insurer, based on Conditions 1 and 2, triggers sufficient demand for coverage that it is economically feasible to market the particular type of coverage.

Condition 1: Identifying the Risk

To satisfy the first condition, estimates must be made of the frequency at which specific events occur and the extent of losses likely to be incurred. Such estimates can be based on data from previous events, or they can draw upon scientific analyses as to what is likely to occur in the future. One way to reflect what experts know and do not know about a particular risk is to construct an exceedance probability (EP) curve.

An EP curve depicts the annual probability that a certain level of losses from a series of different events will exceed a certain magnitude. The loss can be reflected in terms of dollars of damage, fatalities, illness, or some other measure. To illustrate with a specific example, suppose an insurer were interested in constructing an EP curve for dollar losses from a catastrophic chemical accident. Using probabilistic risk assessment, the insurer would determine the likelihood of different accidents occurring and their respective losses. It then would determine the resulting probabilities of exceeding losses of different magnitudes. The uncertainty in the probability of an event occurring and the magnitude of dollar losses as reflected in the 5 percent and 95 percent confidence interval curves are depicted in Figure 6-1.

When dealing with environmental and safety risks, the key question that needs to be addressed when constructing an EP curve is the degree of uncertainty with respect to the probability of the event and the resulting consequences. If the losses are associated with low-probability events, then, by definition, there will be limited past data to estimate these risks. In such a case, one may have to rely on scientifically based estimates by experts about how likely these events are and the impacts they will have on employees and those outside the firm who may be affected.

Condition 2: Setting Premiums for Specific Risks

Once the risk has been identified, the insurer needs to determine what premium it can charge to make a profit while not subjecting itself to an unacceptably high chance of a catastrophic loss. There are a number of factors that play a role in determining what prices companies would like to charge.

Ambiguity of Risk. Not surprisingly, the higher the uncertainty about the probability of a specific loss and its magnitude, the higher the premium will be. To better understand the premium-setting process in these situations, Kunreuther et al. (1995) conducted a survey of 896 underwriters in 190 randomly chosen insurance companies to determine what premiums would be required to insure a factory against property damage from different types of events (e.g., underground storage tank damage or a severe earthquake). The survey results examined changes in pricing strategy as a function of the degree of uncertainty

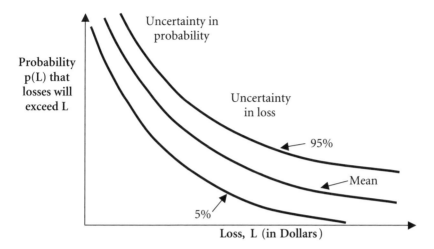

FIGURE 6-1. Example of Loss Exceedance Probability Curves

in either the probability or loss (or both). For highly ambiguous risks, the average premiums were between 1.5 and 1.8 times higher than risks where the probability and losses were well specified.

Adverse Selection. If the insurer sets a premium based on the average probability of a loss using its entire portfolio of policyholders protected against a particular hazard as a basis for this estimate, those at the highest risk will be the most likely to purchase coverage.[2] In an extreme case, the poor risks will be the only purchasers of coverage, and the insurer will lose money on each policy sold. This situation, referred to as *adverse selection*, occurs when the insurer cannot distinguish between the probabilities of a loss for good- and poor-risk categories. The assumption underlying adverse selection is that purchasers of insurance have an informational advantage by knowing their risk type.

Insurers, on the other hand, must invest considerable expense to distinguish among risks. Mandatory insurance is one way to lessen the adverse selection problem, although it can result in good-risk insurance buyers bearing some of the costs of poor-risk insurance buyers. Inspections can lessen the problem because they enable insurers to gather the information they need to distinguish good- and poor-risk categories and set premiums that reflect these differences.

Moral Hazard. Providing insurance protection may lead those who are protected financially to behave more carelessly than before they had coverage, thus increasing the probability of a loss.[3] If the insurer cannot predict this behavior and relies on past loss data from uninsured individuals to estimate rates, the resulting premium is likely to be too low to cover losses.

One way to avoid moral hazard is to introduce *deductibles* and *coinsurance* as part of the insurance contract. A large deductible can act as an incentive for those insured to continue to behave carefully after purchasing coverage because they will be forced to cover a significant portion of their loss themselves. With coinsurance, the insurer and the insured share the loss together. An 80 percent coinsurance clause in an insurance policy means that the insurer pays 80 percent of the loss (above a deductible), and the insured pays the other 20 percent. As with a deductible, this type of risk-sharing arrangement encourages safer behavior because those insured want to avoid having to pay for some of the losses.[4] Another way of encouraging safer behavior is to place upper limits on the amount of coverage an individual or enterprise can purchase so they have to incur some of the costs themselves if they suffer a loss.

Correlated Risk. *Correlated risk* refers to the simultaneous occurrence of many losses from a single event. If a risk-averse insurer faces highly correlated losses from one event, it may want to set a high enough premium not only to cover its expected losses but also to protect itself against the possibility of experiencing catastrophic losses.

Condition 3: Sufficient Demand for Coverage

In theory, insurers can offer protection against any risk that they can identify and for which they can obtain information to estimate the frequency and magnitude of potential losses. However, because of problems of ambiguity, adverse selection, moral hazard, and highly correlated losses, they may want to charge premiums that considerably exceed the expected loss. For some risks, the desired premium may be so high that there would be very little demand for coverage at that rate. In such cases, the insurer will not invest the time and money to develop the product.

More specifically, the insurer must be convinced that there is sufficient demand to cover the development and marketing costs of the coverage through future premiums received. If an insurer's portfolio leaves it vulnerable to the possibility of extremely large losses from a given disaster—because of adverse selection, moral hazard, or a high correlation of risks—then the insurer will want to reduce the number of policies in force. If there are regulatory restrictions that limit the price insurers can charge for certain types of coverage, then insurers may not want to provide protection against these risks at all because the premiums they can charge will not be sufficient to cover the risks plus administrative and marketing costs.

Given these three insurability conditions, an insurer is most likely to want to provide coverage against a given risk when (1) an EP curve can be constructed from historical data about harmful incidents and precursor events, together with estimates by scientists and engineers on the likelihood of future accidents; and (2) the premiums are affordable for insurance buyers. When these conditions are met, insurance is likely to be an attractive complement to environmental and safety regulations. This requirement for insurability suggests that the most promising areas for combining mandatory insurance with regulation are those risks for which there are considerable data from past events or sound estimates by scientists and engineers as to the likelihood and consequences of future events. These circumstances allow the insurance company and government to construct an EP curve.

Demand for Insurance and Risk-Reducing Management

Even when insurance against environmental and safety risks is available, prospective buyers may not want to purchase coverage unless it is required. They also may not want to invest in risk-reducing management practices or other measures because they misperceive the risks and have short-term horizons when evaluating the expected benefits of these investments. This section explains why buyers may not be interested in insurance coverage and protective measures, and provides a rationale for a private-public partnership that combines mandatory insurance and regulation.

Lack of Interest in Voluntary Insurance

There was a period in the 1980s when insurers refused to offer environmental coverage. In more recent years, coverage for some types of environmental and safety risks has been available, but few firms have bought policies. Buyer interest in private insurance among underground storage tank owners and firms facing chemical risks, for example, has been too weak to support a voluntary insurance market.

Underground Storage Tank Coverage. Large firms often self-insure and purchase coverage only against catastrophic losses. Smaller firms face budget constraints and have limited assets, so they view insurance as unattractive. This was the principal reason why private insurers were not able to market insurance for underground storage tanks (USTs). In the 1984 amendments to the Resource Conservation and Recovery Act (RCRA), firms with USTs were required to show financial responsibility through insurance or some other form of collateral. Small firms claimed that they could not afford coverage and the EPA established state guarantee funds, financed primarily through gasoline sales taxes, to provide tank owners with the coverage required by RCRA (Boyd and Kunreuther 1997).

Insurance against Chemical-Related Risks. Recent proposals for a voluntary insurance program for dealing with chemical-related risk have not taken hold despite a cooperative effort by the insurance industry, government, and academia to implement such a program (Collins et al. 2002). Several years ago, a task force convened by the Wharton Risk Management and Decision Processes Center initiated two pilot studies using private inspectors and insurance. The task force consisted of the EPA's Chemical Emergency Preparedness and Prevention Office (CEPPO), EPA Region III, and the State of Delaware's Department of Natural Resources and Environmental Control (DNREC) (McNulty et al. 1999; U.S. EPA 2001).

In these pilot studies in Delaware and Pennsylvania, private auditors examined facilities' risk management plans and helped ensure compliance with Section 112(r) of the CAAA at those water chlorination and ammonia refrigeration facilities that had purchased insurance coverage. Ammonia and chlorine were the chemicals selected in the experiment because 80 percent of the sites required to report under Section 112(r) of the CAAA use those chemicals, and the task force was confident that the auditors could be trained to conduct audits in chlorine and ammonia facilities in a two-day training period.

In Delaware and Pennsylvania, the owners and operators of facilities were sympathetic to having third-party inspections if they received certain benefits in return. Facility owners said they would be especially interested if the EPA or a regulatory agency gave them a seal of approval based on the results of the inspection, and if they had the promise of receiving lower premiums on their accident insurance policies if the inspection showed they were operating in a safe manner. To date, the program has not moved forward, in part because the insurance is voluntary and in part because of the ambiguity associated with the risk, making it difficult for insurers to offer premium discounts.

Why Voluntary Insurance Is Unattractive

In addition to the desire for large firms to self-insure and small firms to allocate their limited resources elsewhere, there are other reasons why firms may not want to purchase insurance coverage voluntarily. Industrial plants may be liable for only a portion of the costs associated with environmental and safety risks, so they do not feel the same need to purchase coverage as they would if they had full responsibility for all the losses associated with a particular event. For example, suppose an industrial plant decreases property values of homes in the surrounding area or disrupts community life because of an accident. The firm causing the accident will very likely not bear the costs for these losses. Ashford and Stone (1991) have reported that for every $1 of direct costs associated with an accident, there are $4 to $10 in social costs not borne by the firm.

In addition, managers frequently underestimate or ignore loss probabilities, so they perceive the insurance rates to be too high relative to their own estimates of the risk. With respect to low-probability events, managers use very simple decision rules and often perceive a loss as too unlikely to happen to them. Studies have also shown that managers rarely even ask for probability information when making choices under uncertainty. In a survey by Huber et al. (1997), not a single person asked for precise probability information. Another group of respondents were given precise probability information, but fewer than one in five of these individuals mentioned the probability when asked what factors influenced their decisions (Huber et al. 1997). Other reasons for the weak demand for voluntary insurance include inadequate assets and short time horizons. These reasons are discussed more fully in our subsequent section explaining why firms tend to underinvest in risk reduction measures.

Private Market Forces that Promote Environmental Insurance

Insurance for environmental risks has emerged in recent years because of private market pressures that have created a significant demand for coverage. Two examples are asbestos abatement coverage and property transfer liability insurance. In both cases, the demand for coverage came about because of a concern by the responsible parties for financial assurance against future damage or losses. Those companies potentially causing the environmental problems did not seek insurance coverage until they were required to purchase it as a condition for doing business. [5]

Asbestos Abatement Coverage. Beginning in the early 1980s, significant pressure developed in the United States to remove asbestos from existing structures. This pressure was fueled by public concern over the health effects of asbestos exposure, as well as by government requirements that schools be surveyed to determine the extent of asbestos contained in their structures. In the early to mid-1980s, both the EPA and OSHA issued regulations for asbestos removal work. These governmental regulations contain clear specifications for job site monitoring, worker protection procedures, transportation and disposal of asbestos fibers, and proper removal techniques.

Relying on this regulatory framework, insurance firms assessed the risks associated with removing asbestos and began offering insurance coverage for contractors doing this work. These insurance policies assumed the liabilities of contractors for property damage and bodily injury created as a result of a release of asbestos fibers in excess of the permissible exposure levels at a job site location. Once insurance became available for contractors performing asbestos removal work, those buying asbestos abatement services demanded that their contractors purchase the coverage. In fact, such insurance became both a license and a prerequisite for doing business.

Property Transfer Liability Insurance. Property transfer liability insurance coverage protects new property owners from having to pay unexpected costs to clean up a previously contaminated site. Prior to purchasing property, lenders perform pre-acquisition site assessments to identify potential contamination. Site assessments, however, have significant limitations in detecting all contamination, leaving the new owner and the lender exposed to potentially significant liability. The insurance coverage protects commercial real estate purchasers and their lenders from liability for contamination that is present, but as yet undetected, on the property. Lenders normally require an inspection and this type of insurance on transferred property.

Lack of Interest in Risk-Reducing Measures

If given free choice, firms may also miss opportunities to enhance their own welfare because they underestimate the risks or benefits from undertaking risk-reducing measures.[6] Consider the case where a chemical plant is deciding whether it wants to invest in a protective measure at a cost of C to reduce the probability of an accident from a high probability (p_h) to a low probability (p_l). The loss, should an accident occur, would be L. Suppose the plant were risk neutral and that its objective was to maximize expected profits. It determined that, by implementing the protective measure, its expected annual savings from taking this action would be $S = (p_h - p_l)L$. If the plant is assumed to achieve S at the beginning of each year and expected to be in business for T years, then the expected discounted benefits of undertaking the protective measure is given by:

$$B = \sum_{t=1}^{T} S / (1+r)^{t-1} \tag{6-1}$$

where r is the annual discount rate. The decision rule facing the chemical plant is a simple one. If $B > C$ then it should invest in the protective measure; otherwise it should not. However, firms do not always follow this simple decision rule for several reasons.

Threshold Models of Choice. For one thing, it may be very difficult for the firm to compute the reduction in probabilities and losses that will occur if it implements such a plan. Instead of undertaking the costs of such an exercise, the plant may estimate that the probability (p_h) of a catastrophic accident under its current operations is below some threshold probability of concern (p^*). It then utilizes the following simplified decision rule: if $p_h < p^*$, do not invest in protection. In other words, the plant would not even compute the expected benefits (B) implied by (6-1) and compare them with the costs of protection.

There is considerable empirical evidence that firms use this type of threshold model for making their decisions about ways that they can reduce their risks. Because Union Carbide perceived the chances of a Bhopal-like disaster as too small to warrant consideration, it did not worry about storing large quantities of methyl isocyanate (MIC). This opened up the possibility of a large chemical release, which occurred when water entered the tank (Shrivastava 1987). Had Union Carbide more accurately accounted for the probability of this catastrophic risk in its decisionmaking, it would have undoubtedly chosen a different method for using and storing MIC.

Short Time Horizons. Another reason why firms may not invest in protection is that managers are guided by short-run considerations because their performance is evaluated annually or, sometimes, even more frequently. Hence, they want to make sure the expected benefits over a relatively short time horizon exceed the upfront investment costs, C, of the protective measure. By truncating the time horizon T, the expected benefits of the protective measure decrease as indicated by (6-1).

To illustrate this point, consider the case where $L = \$6,000,000$, $p_h = 1/100$, and $p_l = 1/200$, so that the expected *annual* benefits from investing in protection is:

$$S = (p_h - p_l)L = (1/100 - 1/200)\ \$6,000,000 = \$30,000$$

If the annual discount rate is 8 percent and the time horizon T were 20 years, then B from (6-1) would be approximately \$318,000. If the managers were using a time horizon of $T = 2$, then B would be only about \$58,000. There would thus be a wide dollar range for C over which the plant managers would want to invest in protection had they used a long time horizon, but would prefer not to incur these costs if they were myopic in their thinking.

Limited Assets. Small firms may also underinvest in protection because they have limited assets (A). Shavell (1986) noted that a firm cannot be fined more than its net worth and used the term *judgment proof* to describe this situation. Rather than investing some of their limited capital in protection, small firms may prefer to take their chances in these situations (Boyd and Ingberman 1994). Ringlieb and Wiggins (1992) show that large firms may form independent subsidiaries with small asset bases to protect the net worth of the parent company from environmental risk. These small units may have no incentive to invest their limited resources in protection. In the above example, suppose that by a spinoff A is reduced to \$3 million so that $S = \$15,000$ and the discounted total benefits (B) from (6-1) were then only \$159,000. In this case, if C were between \$159,000 and \$318,000, the firm would not implement protective measures because of its limited assets, although it would have taken this step had $A \geq \$6$ million.

Interdependencies in Multidivision Firms. In decentralized firms, the economic incentive for any division in an organization to invest in risk-reduction

requirements, it might have an interest in sharing that information with the relevant government agency to convince it to adjust the rules.

The early experience of the Hartford Steam Boiler Inspection and Insurance Company (HSB) suggests how this might work (Weaver and McNulty 1991). HSB initiated boiler inspections as a condition for insurance in the late 1860s. It initiated this program because of enormous public concern regarding the safety of boilers after a major explosion on the steamboat *Sultana* on the Mississippi River in 1865, which killed 1,238 passengers. The cost of engineering and inspection services made up a large part of its insurance premium. In an effort to reduce future risks, HSB also undertook studies of boiler construction, which eventually led boilermakers to adopt safer designs.

The government also learned from the experience of HSB and other insurers. Today, one of the key elements leading to the reduction in the number of boiler accidents is that almost all the states in the United States require annual inspections of pressure vessels by a representative licensed by the state, county, or municipality in which the facility is located.[8] Inspectors are qualified by either a formal examination or a certificate of competency issued by the National Board of Boiler and Pressure Vessel Inspectors. Firms that operate boilers normally purchase insurance with rates determined by the inspection process. Steam boiler accidents have been very rare ever since these requirements were put in place.

Ideally, one would propose a voluntary insurance program similar to the one adopted by HSB. Because of the reasons noted earlier, however, there is unlikely to be sufficient interest in coverage for this to be successful in all areas, especially among small firms or where adverse selection is likely. Mandatory insurance can address these problems.

Insurers will not necessarily find mandatory insurance attractive. The insurance industry does not want to be viewed as a police officer whose job it is to enforce a particular regulation. Insurance firms may also believe they cannot accurately set rates based on risk because of a lack of historical data and good scientific information. Firms' lack of interest in marketing terrorism coverage after September 11th led to the passage of the Terrorism Risk Insurance Act (TRIA) of 2002, whereby the federal government now plays a role in providing protection against catastrophic losses from terrorism. If insurers are not willing to provide coverage for some firms, there is a need for risk pools or residual market mechanisms to cover these risks.

Benefits of Pairing Inspections with Insurance

This section considers more fully why a system of mandatory insurance coupled with increased private inspections can complement current governmental inspection programs. Regulation combined with insurance could be a way to obtain better information to rate risks, to reward firms for undertaking risk-reducing measures, and to disclose this information to the public. This section also looks at possible variations in the characteristics of insurance actions likely to make the partnership more productive.

Insurance is likely to have greater risk-reducing potential if insurers include inspections, along with other forms of risk assessment, as part of the insurance-rating package.[9] Private insurance inspections can play an important role for several reasons. At the most basic level, insured firms will be more aware of environmental and safety risks as well as regulatory obligations. This promises to be especially valuable in areas of health, safety, and the environment that are plagued by low inspection levels.

Gathering Risk Information

Inspections enable the insurer to determine what the losses are likely to be as a function of the investments by firms in risk-reducing measures, such as the adoption of an environmental or total quality management system. Insurers can also provide guidance to the firm about the types of management practices likely to be most profitable for assuring compliance with regulations and reducing costs associated with environmental risks. If insurers increase their inspections of a firm's safety practices prior to policy renewals, firms will have incentives to comply with the regulations and take environmental risk management seriously.

Use of Claims Data to Modify Existing Standards

Studying information about claims, incidents, and noncompliance may identify recurring events and high-cost problems needing new laws or standards. If an insurer has a large enough set of clients and can pool information so as not to reveal identities of firms, then it can provide valuable information to the public sector on the types of claims that have been made. This will enable the public sector agency to modify codes and standards in an appropriate fashion.

Industrywide associations, such as the National Fire Protection Association, can also collect claims data from a number of insurers, sanitize the information, and provide the data to industry and public sector agencies to develop model codes. These model codes then get incorporated into legal standards and regulations (e.g., requirements for sprinklers and fire-detection systems in certain buildings). In some agencies, such as the National Highway Traffic Safety Administration, government and insurers study incident data, identify primary contributors to risky and costly occurrences, and shape new guidelines or standards to reduce accident rates (Metzenbaum 2003a).

Insurers have an actuarial, evidence-based culture. With a few noteworthy exceptions, regulators do not, tending to focus instead on individual cases (Metzenbaum 2003b). Pairing insurance with regulation should allow the two disparate cultures to tap each other's strengths.

Rewarding Firms for Reducing Risks

Insurers always have the option of raising rates to reflect additional risks they uncover. They can also refuse to provide coverage to firms seen as too risky, either because those firms have not complied with a set of regulations, have failed to adopt a set of best practices, or have had a bad risk record. Thus insurance review is likely to carry great weight with the firm.

Insurers can also bestow reputation-enhancing rewards on firms operating at the highest level of compliance and taking risk-reducing actions beyond their formal obligations. Seals of approval or gold stars are valuable to the firm to the extent that customers, employees, or investors make decisions on the basis of safety and environmental records of different organizations. Some commercial partners will see the gold star as the designation of a quality operation and favor doing business with these firms. For example, Ford and GM require their suppliers to be ISO 14001 certified, as discussed by Andrews et al. in Chapter 5.

The firm that earns the seal or gold star will have an incentive to reveal its third-party commendation to the public as well as to the government. Regulatory agencies can use this information to target inspections to firms that do not have this official recognition. By reducing the universe of firms for possible inspection, there is a greater chance that those who have not complied with the regulation will be audited by a governmental agency. By raising the probability of a public inspection, more firms should adhere to regulations over time.

An insurance commendation is likely to have greater veracity than other sorts of third-party certifications because most third-party inspectors are paid a fee for their services by the inspected firm, and therefore feel a constant tug to keep their customer happy without a strong counterbalancing financial tug to identify risks that may require costly corrections. Insurers, in contrast, have a direct financial interest in reducing risk through their inspections.

Identifying Parties Operating without Legal Permits

Mandatory insurance does not ensure that everyone will actually comply with the mandate and purchase the insurance. Owners of the Rhode Island night club where 99 people tragically died in a fire in 2003, for example, had neglected to purchase their mandated workers' compensation insurance, leading the state to slap a million dollar fine on them after the fire (*Boston Globe* 2003).

It is reasonable to believe that noninsured firms are also more likely to be ones that have not adhered to specific government requirements, such as filing for a permit to operate. The "nonfiler" problem can be significant, especially for small businesses. When Massachusetts made an effort to locate all facilities in three small business sectors that should have been regulated, it found 2,200 facilities. Before this concerted effort, the state estimated it only had 380 firms in the state that met the criteria for regulation (April and Greiner 2000). Similarly, in its efforts to clean up the Charles River in Massachusetts, the EPA discovered numerous sources discharging wastewater to the river without a permit (Metzenbaum 2002). Unfortunately, given lack of inspection and enforcement personnel, this type of detective work by government agencies is not the norm.

Insurance companies, however, may have a reason to make this sort of detective work their norm. They have a financial incentive to find nonfilers—nonfilers are potential customers. Government could also make insurers' identification processes easier with a few complementary actions of its own. A mandatory insurance program might be structured to require all firms to register their insurance status to make it easier to identify noninsured firms. It could require every employer to post a certificate of insurance prominently in the workplace and at the street entrance where the public can see it.

The government requirement could include a format for posting the certification with the phone number and Web site of the government office where information about the requirement can be obtained or complaints registered. Government could also advertise or otherwise raise worker awareness of the insurance-holding requirement in locations where employees are likely to read it. In addition, it could require companies to file a copy of their certificate of insurance annually and use these filings to create an online database listing firms that hold insurance certification.

At the same time, government could work with insurers to encourage major buyers to pay attention to their suppliers' insured status. Government might do this by awarding positive publicity to buyers that require all suppliers to have insurance, or it might even require buyers to confirm insurance ownership for purchases in excess of some minimum threshold. This requirement for confirmation of insurance ownership among suppliers could also apply to government as a buyer.

A Proposed Pilot Study in Louisiana

The concepts discussed in the previous section draw on a set of discussions with industrial firms, insurers, and public sector agencies over the past five years through a series of roundtables at the Wharton Risk Management and Decision Processes Center on ways of integrating insurance, inspections, and regulations for reducing environmental and safety risks. Here we propose a strategy for further study of these questions.

A pilot study in the state of Louisiana has been proposed by the Louisiana

Workers' Compensation Corporation (LWCC) to explore how a private insurer might adjust its behavior relative to insured parties and the government in order to achieve greater levels of risk reduction and regulatory compliance.[10] The study would also determine the value and feasibility of adjustments in the practices of government agencies.

This project is being explored because workers' compensation has long been an area where government and insurers have worked closely together. The first constitutional workers' compensation law was passed in Wisconsin in 1911.[11] By 1920, 42 states had such laws; Mississippi, in 1949, was the last state to join the program (Welch 1994).

Workers' compensation insurance arose in response to concerns about the large litigation costs associated with tort law. Tort law required workers to show fault in order to obtain compensation from their employer. State legislators, concerned about the undue burden being placed on occupationally injured workers, mandated financial assurance by companies to cover compensation costs. Workers' compensation insurance is a no-fault mechanism where the only evidence required for a claim is that the accident or injury be job-related (Viscusi 1991).

In Louisiana, the insurance program covers almost 30 percent of workers in the state, with the other 70 percent covered through other financial assurance mechanisms (Yuspeh 2003). The program provides cash benefits and medical care for work-related injuries and diseases. Since premiums for many employers are linked by experience-rating formulas to benefits paid to their employees, firms have financial incentives to reduce risk levels.

There is debate about whether workers' compensation has achieved its intended improvement in safety. Mont et al. (2001) conclude that the evidence that experience rating improves safety is only mildly persuasive. This may be because, as benefits increase as they did after 1972, the cost of an injury to the workers decreases, so they may have a tendency to be less attentive to the hazards of their job (i.e., the moral hazard problem). Of course, inattentiveness is unlikely to be a problem when workers know they may experience debilitating physical pain from an injury.

The opposing argument is that higher insurance premiums put pressure on employers to have a safer workplace. Viscusi (1991) contends that if the safety incentives of workers' compensation were removed, fatality rates would rise over 30 percent in the United States. This translates into an increase of 1,200 workers who would die from job-related accidents. On balance it appears that workers' compensation has been a major success in relieving the burden on injured workers of initiating a tort process biased against the worker to recover damages caused by an accident, although some state systems face serious cost-control problems.

In 1983, the Louisiana Office of Workers' Compensation Administration (OWCA) was created to resolve disputes and oversee the workers' compensation system, but it was not until the late 1980s that dispute resolution moved from the district courts to OWCA. In 1992, the legislature created the LWCC, a private insurer with initial reserves backed by the full faith and credit of the state. The LWCC competed with other insurers in the state and was also the insurer of last resort.

The LWCC is no longer dependent upon the state for the backing of its reserves, which have grown to almost $300 million. The LWCC is proactive in the area of risk reduction. It feels that encouraging firms to move beyond compliance with OSHA standards will enhance Louisiana's reputation as a preeminently safe place to work, encouraging the relocation of industries and people to within its borders.

Nature of Proposed Pilot Program

A pilot program would address businesses that employ workers in specific high-severity/low-frequency workers' compensation classes. For these businesses, safety should be an especially important matter because workers' compensation rates are particularly high because of perceived high risks that are difficult to predict. Management of extreme risk is the key factor affecting reinsurance rates and thereby primary insurance rates. It thus represents a major challenge to the workers' compensation industry. If LWCC can identify key precursor events and other risk factors, it may be able to reduce risk levels.

Such a study could address the following questions:

- If LWCC undertakes an expanded inspection regime of firms in high-severity/low-frequency industries, will it be able to identify precursor conditions (key risk factors) that will help it reduce risks and accidents?
- Can government use relevant information from LWCC inspections to improve its regulatory requirements in order to focus more accurately on improved management and other risk-reduction opportunities, possibly dropping some regulatory obligations not highly correlated with risk?
- Will LWCC be willing to share its inspection information (raw data or analyses) with government? Will it be interested in recommending improvements to government regulations, such as adding or eliminating legal requirements?
- Will insured parties be willing to disclose their insured status to the government and the public? Will government or the insured find this information useful? How and for what purpose?
- Will insured parties welcome high-performance ratings as an opportunity to win recognition for performance beyond compliance? Why? Will LWCC be willing to award high-performance ratings to its customers?
- Will OSHA want to adjust its inspection protocols to take advantage of changed insurer behavior? What sort of adjustments will it want to make?

Will it change its enforcement scheduling?
- Finally, and perhaps most important, will the level of accidents, harm, or injury decline after these changes are introduced?

The study could be designed in various ways. In its simplest form, the information could be tracked longitudinally to try to detect correlated changes. If there are enough firms within the same or similar primary class codes, it may be possible to divide the group randomly into a treatment group participating in the expanded inspection regime and a control group. Accident, injury, and absentee data could be studied routinely to determine if the expanded inspection regime with its special features reduces worker risks, and if it produced data that helped the LWCC identify previously unknown risk factors as they relate to extreme risk.

It may be possible to design the insurance inspections as an extension and enhancement of OWCA's Consultation Program. Through this OSHA-funded program, employers can find out about potential hazards at their worksites, improve their occupational safety and health management systems, and even qualify for a one-year exemption from routine OSHA inspections.

Anticipated Benefits from Proposed Program

For the LWCC and other insurers, reduction in severe accidents could lead to higher profitability and lower premiums because of improved safety performance in high-severity classes. Firms that have fewer accidents will be rewarded with lower insurance rates. They will also reduce ancillary accident costs, which are estimated to be as much as eight times the direct accident costs. Firms might also enjoy some level of regulatory relief as well as increased business from larger customers concerned about their suppliers' workplace safety.

Workers would enjoy the obvious benefits of a safer work environment and especially an environment that focuses on preventing extreme events. The program could look at the potential benefits of requiring the posting of a business's specific safety performance. It could serve to increase employee morale and help the firm attract a more capable work pool.

The insurance company would enjoy a much deeper and detailed knowledge of the safety culture and structure of the businesses it inspected under this program. This would enable it to do a better job underwriting and differentiating the risks within high severity class codes. The enhanced knowledge obtained through the inspections should add to its repertoire of risk reduction techniques, resulting in better overall industry safety.

OSHA has limited resources for making compliance inspections. The use of voluntary third-party inspections by the insurance industry could become the centerpiece of OSHA's compliance approach to enhancing workplace safety. OSHA inspectors could instead concentrate the agency's resources on enforcement. OWCA could, in effect, have its Consultation Program expanded and Louisiana could have a larger number of businesses geared to enhanced workplace safety.

Conclusions and Future Research Directions

This chapter explores the potential value of mandatory insurance coupled with inspections and disclosure as a complement to government regulation. By linking inspections with a required insurance policy and setting insurance rates based on risk, both large and small firms will realize that it is worthwhile for them not only to meet a given rule but also to adopt protective measures that will benefit both them and the public. Such a program will be particularly attractive to a firm if the inspector can provide special risk management services in addition to its audit function.

Mandatory insurance coupled with private and government inspections has the potential to reduce risk more than is feasible under the current regulatory system with its limited enforcement capabilities. Given the limited enforcement personnel in regulatory agencies, it is important to be able to target those firms that have not complied with the insurance requirements for government audits.

Mandatory insurance and third-party inspections are also likely to generate more information about environmental, health, and safety management practices than is currently available. Because the insurance industry has a financial incentive to study the correlation between the inspection information it gathers and the resulting risk, insurers should be able to provide businesses with advice for improving their management and provide government agencies with data that could help them craft more effective regulation.

There is a need to clarify issues and undertake research to better understand the challenges of using private inspections, improving risk assessments, and investigating the role of private voluntary insurance in conjunction with inspections for reducing health, safety, and environmental risks. Three areas needing additional consideration are discussed below.

Challenges in Utilizing Private Inspections

The implicit assumption in our analysis in this chapter is that if a firm undertakes a private inspection and does not want to incur the costs associated with investing in improved management and other protective measures, it is the firm's choice to do so. Furthermore, we have assumed that there is no obligation by firms or inspectors to report the results of the audit to the regulatory authority. In practice, there may need to be exceptions to this assumption. For example, if an inspection team were to find that a hazard at a facility put employees in imminent danger, the inspectors would be faced with a conflict between their specific duty to maintain confidentiality with their client and their general obligation to avoid exposing individuals to possible injury or death.

One way to deal with this situation would be to require inspectors to reveal information about imminent harm to the relevant regulatory authority if the firm refuses to take any remedial action within a predetermined period of time. Such a procedure would be in line with current OSHA policy for consultants to the agency who discover similar situations (OSHA 1989). As pointed out in Collins et al. (2002), more studies need to be undertaken to determine under what situations inspectors have an obligation to reveal the results of their findings to employees, citizens, or the regulatory authority.

There are a number of open issues on the role of inspection that should be explored in future research. These include the ability of inspectors to determine how safe a firm actually is; the asymmetry of information among firms, insurers, and inspectors; and the ability to estimate the risk of accidents and the costs of preventive actions. There is also a need to establish a certification mechanism for selecting and training third-party inspectors, determining what actions need to be taken if an inspector discovers that the firm is operating in ways that may be hazardous to employees or residents in the region, and determining when inspectors are liable. More generally, how can one maintain the confidentiality of a sound private inspection process while also making sure the public and employees are adequately protected?

Improving Risk Assessments

It is much easier to assess relatively high probability risks such as worker safety than it is to assess the risks of extreme events for which there are limited historical data and for which the science is more controversial. Experts are likely to disagree with each other and there is no easy way to settle their differences. There is a need to collect better data with which to determine risk-based premiums. Currently, the Wharton Risk Management and Decision Processes Center is engaging in two types of data collection efforts in this regard with the U.S. Environmental Protection Agency. The first of these data collection efforts uses accident history data from the U.S. chemical industry; the second is concerned with the performance of management systems designed to improve the environmental, health, and safety performance of companies.

Accident history data have been collected on the performance of the U.S. chemical industry for the period 1995–2000, under Section 112(r) of the Clean Air Act Amendments of 1990 (Kleindorfer 2006). Every chemical facility in the United States was required to provide these data for any listed chemical above threshold quantities. Elliott et al. (2003) and Elliott et al. (2004) indicate how these data can be used to analyze potential relationships between the characteristics of the facility, the nature of regulations in force, and the level of community pressure brought on the facility.

These same data can be linked to financial information to analyze the association, if any, between the financial characteristics of the parent company of a facility and the frequency or severity of accidents. Similarly, property damage estimates and associated indirect costs can be used to assess the consequences of environmental, health, and safety incidents on overall company performance and provide valuable insights for insurance underwriting for such accidents. Finally, the same data can be used to assess worst-case consequences from such incidents, including those that might arise from site security risks associated with terrorism.[12]

The second data collection project is a study of precursors in organizations and the systems that have been put into place to report and analyze these data. *Precursors* are defined as incidents that, under different circumstances, could have resulted in major accidents. Linking these data on accident precursors to accident history data may enable one to identify categories of precursors that give early warnings of the potential for major accidents. Audit tools and other aspects of precursor management can then focus not just on emergency response but on the range of prevention and mitigation activities that can help avert major accidents and injuries (Phimister et al. 2003).

An example of how precursor data has been utilized for better estimating risks and reducing insurance premiums comes from Syncrude Canada Ltd., the world's largest producer of crude oil from oil sands. By developing a detailed database based on accidents that almost occurred, Syncrude has shown that incident rates have decreased as near-miss reporting has increased. In 2002, the company earned a $1.6 million end-of-year rebate of workers' compensation insurance premiums, currently paid at the rate of $0.23 per $100 in payroll. By comparison, the roofing industry pays $11 per $100 in payroll to insure its workers (Thompson 2003).

In summary, a system of mandatory insurance coupled with private inspections holds great promise as a way to reduce safety and environmental risks in a self-propelling, self-financing manner. By using the private risk-management systems of insurance to complement understaffed regulatory programs, policymakers can address two problems that threaten the viability of existing regulatory programs. The current system presumes that two key conditions are satisfied: first, that regulated parties comply with regulations and, second, that those regulations, when adopted, do in fact reduce risk. Inadequate program inspection capacity seriously compromises the first condition in many programs, and limited available data make it difficult to evaluate the second condition.

Mandating insurance that would be accompanied by private inspections for all regulated parties could address both problems in a self-financing manner.

Insurance industry loss-prevention personnel could serve as third-party inspectors, identifying—through inspection and claims—practices that reduce risk. They could then price insurance to encourage these practices. The challenge is to field test the approach proposed here, which is easier said than done because testing requires cooperation from government and insurers.

One possible field test involving a government-financed insurer has been outlined in this chapter. The authors hope there will be others willing to test this promising, private-sector-leveraging approach and enter into a research partnership to assess its viability.

Acknowledgments

Our thanks to Max Bazerman, Jay Benforado, Cary Coglianese, and Jennifer Nash for helpful comments on an earlier draft of this chapter. Support for this research under the U.S. Environmental Protection Agency's Cooperative Agreement C R 826583 with the Wharton Risk Management and Decision Processes Center of the University of Pennsylvania is gratefully acknowledged.

Notes

1. For a more detailed discussion of insurability in the context of environmental risks, see Freeman and Kunreuther (1997), Chapter 4.

2. For a survey of adverse selection in insurance markets, see Dionne and Doherty (1992).

3. See Winter (1992) for a survey of the relevant literature on moral hazard in insurance markets.

4. For more details on deductibles and coinsurance in relation to moral hazard, see Pauly (1968).

5. For more details on both types of coverage, see Freeman and Kunreuther (1997).

6. This discussion is based on Kunreuther et al. (2002).

7. The fine may be larger than D/q to the extent that there are impacts such as fear that cannot be measured financially but that the general public feels should be compensated. In practice, environmental agencies base fines on several factors. The Clean Water Act, for example, sets out six factors for setting penalties: the seriousness of the violation(s), the economic benefit resulting from the violation, any history of such violations, any good-faith efforts to comply with the applicable requirements, the economic impact of the penalty on the violator, and such other matters as justice may require (Section 309 (d)).

8. Florida and South Carolina are two states that do not require boiler inspections.

9. Insurers already make inspections routine practice for those facilities with large premiums or loss ratios. The State of Texas mandates insurance inspections for certain classes of facilities (Texas Workers Compensation Commission Rule 166.4).

10. Larry Yuspeh from the Louisiana Workers' Compensation Corporation suggested this study. He has spent considerable time with us and others in Louisiana to craft the details of the proposed pilot program described in this section.

11. New York's law, passed in 1910, was deemed unconstitutional.

12. For recent details on this aspect of the chemical Accident History Database RMP*Info, see Kleindorfer et al. (2003).

References

April, Susan, and Tim Greiner. 2000. Evaluation of the Massachusetts Environmental Results Program. In National Academy of Public Administration, *environment.gov: Transforming Environmental Protection for the 21st Century, Research Paper No. 1, Volume 1.* Washington, DC: NAPA.

Ashford, Nicholas A., and Robert F. Stone. 1991. Liability, Innovation and Safety in the Chemical Industry. In *The Liability Maze,* edited by Peter W. Huber and Robert E. Litan. Washington, DC: Brookings Institution Press, 367–427.

Boston Globe. 2003. Derderians Appeal R.I. Insurance Fine (Associated Press) May 3, B8.

Boyd, James, and Daniel Ingberman. 1994. Non-Compensatory Damages and Potential Insolvency. *Journal of Legal Studies* 23(2): 895–910.

Boyd, James, and Howard Kunreuther. 1997. Retroactive Liability or the Public Purse? *Journal of Regulatory Economics* 11(1): 79–90.

Coglianese, Cary, and David Lazer. 2003. Management-Based Regulation: Prescribing Private Management to Achieve Public Goals. *Law and Society Review* 37(4): 691–730.

Collins, Larry, James C. Belke, Marc Halpern, Ruth A. Katz, Howard C. Kunreuther, and Patrick J. McNulty. 2002. The Insurance Industry as a Qualified Third-Party Auditor. *Professional Safety* 47(4): 31–8.

Dionne, Georges, and Neil Doherty. 1992. Adverse Selection in Insurance Markets: A Selective Survey. In *Contributions to Insurance Economics,* edited by Georges Dionne. Boston: Kluwer Academic Publishers, 97–140.

Elliott, Michael R., Paul R. Kleindorfer, and Robert A. Lowe. 2003. The Role of Hazardousness and Regulatory Practice in the Accidental Release of Chemicals at US Industrial Facilities. *Risk Analysis* 23(5): 883–96.

Elliott, Michael R., Yanlin Wang, Robert A. Lowe, and Paul R. Kleindorfer. 2004. Environmental Justice: Frequency and Severity of US Chemical Industry Accidents and the Socioeconomic Status of Surrounding Communities. *Journal of Epidemiology & Community Health* 58: 24–30.

Freeman, Paul K., and Howard Kunreuther. 1997. *Managing Environmental Risk through Insurance.* Washington, DC: American Enterprise Institute Press (softcover); Boston: Kluwer Academic Publishers (hardcover).

Heyes, Anthony. 1998. Making Things Stick: Enforcement and Compliance. *Oxford Review of Economic Policy* 14: 50–63.

Huber, Oswald, Roman Wider, and Odilo Huber. 1997. Active Information Search and

Complete Information Presentation in Naturalistic Risky Decision Tasks. *Acta Psychologica* 95(1): 15–29.

International Organization for Standardization (ISO) 2004. ISO 9000 and ISO 14000—In Brief. http://www.iso.ch/iso/en/iso9000-14000/index.html. Accessed on January 31, 2005.

Kleindorfer, Paul R. 2006. The Risk Management Program Rule and Management-Based Regulation. In *Leveraging the Private Sector: Management-Based Strategies for Improving Environmental Performance*, edited by Cary Coglianese and Jennifer Nash. Washington, DC: Resources for the Future, Chapter 4.

Kleindorfer, Paul R., James C. Belke, Michael R. Elliott, Kiwan Lee, and Harold Feldman. 2003. Accident Epidemiology and the U.S. Chemical Industry: Accident History and Worst Case Data from RMP*Info. *Risk Analysis* 23(5): 865–81.

Kunreuther, Howard, Jacqueline Meszaros, Robin Hogarth, and Mark Spranca. 1995. Ambiguity and Underwriter Decision Processes. *Journal of Economic Behavior and Organization* 26(3): 337–52.

Kunreuther, Howard, Patrick McNulty, and Yong Kang. 2002. Third-Party Inspection as an Alternative to Command and Control Regulation. *Risk Analysis.* 22(2): 309–18.

Kunreuther, Howard, and Geoffrey Heal. 2005. Interdependencies within an Organization. In *Organizational Encounters with Risk*, edited by Bridget Hutter and Michael Power. Cambridge: Cambridge UP, 190–208.

McNulty, Pat J., Leon C. Schaller, and Robert A. Barrish. 1999. Evaluating the Use of Third Parties to Measure Process Safety Management in Small Firms. Paper presented at 1999 Annual Symposium, Mary Kay O'Connor Process Safety Center, Texas A&M University, October 26, 1999, College Station, TX.

Metzenbaum, Shelley H. 2002. Measurement that Matters: Cleaning Up the Charles River. In *Environmental Governance*, edited by Donald F. Kettl. Washington, DC: Brookings Institution Press, 58–117.

———. 2003a. Strategies for Using State Information: Measuring and Improving Program Performance. Washington, DC: IBM Center for The Business of Government.

———. 2003b. More Nutritious Beans. *Environmental Forum* 20(2): 19–41.

Mont, Daniel, John F. Burton, Jr., Virginia Reno, and Cecili Thompson. 2001. *Workers' Compensation: Benefits, Coverage, and Costs, 1999 New Estimates and 1996–1998 Revisions*. Washington, DC: National Academy of Social Insurance.

Multi-State Working Group on Environmental Performance (MSWG). 2004. External Value Environmental Management System Voluntary Guidance: Gaining Value by Addressing Stakeholder Needs. http://www.mswg.org/documents/guidance04.pdf (accessed January 31, 2005).

National Academy of Public Administration. 2001. Third-Party Auditing of Environmental Management Systems: U.S. Registration Practices for ISO 14001. Washington, DC: NAPA, 30–43.

OSHA (Occupational Safety and Health Administration). 1989. 29 CFR Part 1908, Consultation Agreements, Final Rule, 49 FR 25094, June 19, 1984, as amended at 54 FR 24333, June 7, 1989.

Pauly, Mark. 1968. The Economics of Moral Hazard: Comment. *American Economic Review* 58 (3): 531–37.

Pfaff, Alexander, and Chris Sanchirico. 2000. Environmental Self-Auditing: Setting the Proper Incentives for Discovery and Correction of Environmental Harm. *Journal of Law, Economics and Organization* 16(1): 189–208.

Phimister, James, Ulku Oktem, Paul R. Kleindorfer, and Howard Kunreuther. 2003. Near-Miss Management Systems in the Chemical Process Industry. *Risk Analysis* 23(3): 445–59.

Ringlieb, Al H., and Steven N. Wiggins. 1992. Liability and Large-scale, Long-term Hazards. *Journal of Political Economy* 98(3): 574–95.

Shavell, Steven. 1986. The Judgment Proof Problem. *International Review of Law and Economics* 6 (1): 45–58.

Shrivastava, Paul. 1987. *Bhopal, Anatomy of a Crisis.* New York: Ballinger Publishing Company.

Soltan, Richard. 2003. Personal communication between Richard Soltan, Region 3 Director of OSHA, and the authors, June 27.

Thompson, Phyllis G. 2003. A Review of "Near Miss" Systems across Industrial Domains: Syncrude Case Study. Washington, DC: U.S. Chemical Safety and Hazard Investigation Board.

U.S. EPA (Environmental Protection Agency). 2001. Third Party Audit Pilot Project in the Commonwealth of Pennsylvania, Final Report. Philadelphia, PA: U.S. EPA Region III, February.

Viscusi, W. Kip. 1991. *Reforming Product Liability.* Cambridge: Harvard University Press.

Weaver, Glen, and J. Bard McNulty. 1991. An Evolving Concern: Technology Safety and the Hartford Steam Boiler Inspection and Insurance Company, 1866–1991. Hartford: Hartford Steam Boiler Company.

Weil, David. 1996. If OSHA Is So Bad, Why Is Compliance So Good? *The RAND Journal of Economics* 27(3): 618–40.

Welch, Edward. 1994. *Employer's Guide to Workers' Compensation.* Washington, DC: BNA Books.

Winter, Ralph A. 1992. Moral Hazard and Insurance Contracts. In *Contributions to Insurance Economics,* edited by Georges Dionne. Boston: Kluwer Academic Publishers, 61–96.

Yuspeh, Larry. 2003. Personal communication between Larry Yuseph, Director of Research and Development, Louisiana Workers' Compensation Corporation, and the authors, June 4.

Part III:
Incentives and Pressures

7

The Promise and Limits of Voluntary Management-Based Regulatory Reform

An Analysis of EPA's Strategic Goals Program

Jason Scott Johnston

Although accounting for only a small share of American economy, the metal-finishing industry has had a profound impact on the natural environment. The industry supplies parts that are essential to products ranging from automobiles and household appliances to aircraft and military hardware. At the same time, metal finishing generates wastes from solvents, cleansers, and metal-ion-bearing aqueous solutions used for plating (such as hexavalent chromium, copper, gold, and cadmium). Acid from finishing cleaners is found in metal-finishing wastewater, and solvents from metal-finishing plants may also be emitted into the air or disposed of in solid form. Many hazardous waste sites that were once solid-waste landfills include contaminants from metal-finishing wastes. It is fair to say that metal finishing is an industry whose environmental impact may far exceed its economic significance.

U.S. environmental laws and regulations were drafted with an acute awareness of the metal-finishing industry's potential environmental impact. The industry is subject to regulation under virtually every federal environmental statute. There is, moreover, substantial evidence that the traditional environmental regulatory regime in the United States has indeed succeeded in vastly improving the industry's environmental performance.

The industrial structure of the metal-finishing industry, however, places inherent limits on the ability of government to use traditional regulatory tools to improve the industry's environmental performance. The industry comprises thousands of small firms, many of which are independent job shops that use a wide variety of metals in finishing a wide variety of products. The industry is geographically concentrated in the most heavily industrialized, and heavily polluted, states. Yet, despite economies in being located close to their manufacturing clients, labor and raw materials costs in countries such as China are so low relative to U.S. levels that metal finishing is increasingly being shifted outside the United States. Metal finishers have faced increasingly intense foreign competition, and a significant fraction of firms operate on relatively thin profit margins and are not highly capitalized.

For purposes of regulatory policy, the significance of metal finishing is not that it is an industry where command and control environmental regulation failed. The evidence, discussed in more detail below, is that command and control regulation has driven metal finishers to either dramatically improve their environmental performance, or else shut down. Metal finishing is significant because the industry squarely raises the fundamental regulatory challenge of how to create incentives for continuing, yet increasingly costly, environmental improvement in an industry comprising predominantly small- and medium-sized, thinly capitalized firms. Underlying the challenges posed by the metal-finishing industry is an even more general and fundamental question: Whether the achievement of America's increasingly ambitious and increasingly expensive twenty-first century environmental goals means the end of small- and medium-sized independent domestic manufacturing firms and their replacement by foreign outsourcing and huge, vertically integrated domestic manufacturing enterprises.

The Strategic Goals Program (SGP) was a voluntary regulatory program initiated by U.S. Environmental Protection Agency (EPA) in 1998 in an attempt to improve the environmental performance of a remaining domestic bastion of small- and medium-sized enterprises (SMEs) in metal finishing. Negotiated by EPA, metal-finishing trade associations, and other stakeholders, the SGP was a direct outgrowth of the Clinton Administration's Common Sense Initiative. Its stated goal was to encourage firms in the metal-finishing sector to go "beyond compliance" with existing regulatory requirements. The metal-finishing SGP set concrete pollution-prevention goals for its participating firms. Among these goals were a 50 percent reduction in water use, a 90 percent reduction in organics emissions, and a 50 percent reduction in metals emissions, all to be met by a target date of 2002.

By setting broad, voluntary goals, EPA sought to direct the attention of managers of metal-finishing facilities toward finding ways to improve their environmental performance. Facilities that participated received public recognition from EPA and could qualify for various kinds of regulatory relief granted by the publicly owned treatment works (POTWs) that are both the primary recipients and regulators of metal finishers' waterborne wastes. Through these incentives, EPA hoped to encourage metal-finishing facilities to improve their environmental management.

The SGP ended in 2003. Now, with a few years' perspective, is an ideal time to step back and analyze this program. I begin this chapter with an overview of the metal-finishing industry's structure as well as the federal environmental regulations that apply to the industry, which are key to understanding the forces behind the SGP. I then describe what is known about the SGP's performance by examining publicly available federal and state data. Next I consider how the SGP's attractiveness and relative success varied by firm type. Finally, I step back and ask whether the metal-finishing SGP is something that EPA can or should attempt to replicate in other industry sectors.

My answer to this final question is "no." The metal-finishing SGP appears to have disseminated information about pollution prevention and encouraged the most significant metal-finishing regulators—POTWs—to reward with regulatory flexibility those firms that undertook to prevent pollution. Its technical assistance component clearly generated some win-win success stories—cases in which relatively uninformed small metal finishers learned how they could both increase their profits and reduce their pollution by being more efficient in their use of water, metals, and other raw material inputs. However, in a highly regulated industry such as metal finishing, where larger firms that have managed to survive have done so only by complying with quite extensive and costly technology-based effluent and emission standards, going beyond compliance is quite costly. The dollar benefits to such firms of marginal regulatory incentives (such as reducing self-sampling requirements) are typically small relative to the cost of additional pollution prevention. Managers of middle-tier metal-finishing firms would perhaps have liked to pursue more extensive pollution prevention measures, but faced capital market constraints that prevented them from doing so. The SGP did not provide the necessary financial assistance to cover this gap. Finally, the SGP did not extend regulators' reach to cover the most egregiously noncomplying firms in the industry—firms that either escape regulatory detection or are not regulated because they discharge into very small POTWs.[1]

Overall the SGP was neither especially ambitious in its goals nor clearly successful in achieving them. Significantly, the SGP failed even to generate the sort of firm-specific database that is necessary to evaluate why firms did or did not participate and whether the initiative caused changes in participants' environmental performance. If EPA wants its experiments in regulatory reform to be more than *ad hoc* curiosities, it must generate reliable, facility-specific quantitative data.

More fundamentally, metal-finishing job shops face economic limitations in their ability to make further improvements in their environmental performance. EPA puts metal finishing that takes place within large company manufacturing facilities (*captive shops*) in a different regulatory category from metal finishing undertaken by independent operators (*job shops*). The SGP was intended to help small, capital-constrained job shops, not large captive shops owned by capital-rich firms such as General Electric and Boeing. From the point of view of the environment and human health, however, what matters is the industrial process, not where it is taking place. Metal finishing is a process with a very high potential for harmful environmental impact.

American society will continue to demand declines in that impact. Yet the experience of the SGP shows that the burden of reducing environmental harm is often too much for small firms to finance. With the advent of global market competition, the economic pressures facing small- and medium-sized American firms have only increased. American society ultimately faces a tradeoff between the preservation of small- and medium-sized domestic manufacturing firms and continuing pollution reduction. If we are to continually move down the curve of diminishing marginal benefit and increasing marginal cost from pollution reduction, then we will lose small- and medium-sized domestic manufacturing.

In metal finishing, significant further improvement in the industry's environmental performance is likely to be achieved only if firms become larger, both through large manufacturers taking more metal finishing in-house (more captives) and through the consolidation of remaining independent job shops. If significant environmental improvements are indeed the goal, then EPA should consider easing the industrial transition they require by adopting a tradable permit scheme such as the one that the agency successfully used in eliminating lead additives from gasoline (Nichols 1997). Such transitioning would prolong the life of small job shops by allowing them to buy emission and effluent credits. EPA's time and resources would be better spent in exploring the feasibility of such a tradable rights regime for metal finishing and similarly structured industries than in attempting more programs like the SGP.

The Structure and Regulatory History of the Metal-Finishing Industry

The metal-finishing industry's structure and regulatory history provide an important backdrop to the SGP. EPA recognized that, although traditional environmental regulation had improved the environmental performance of the metal-finishing industry, imposing further regulatory controls would inflict substantial economic hardships on this already financially stressed industry and cause many small shops to go out of business.

Industry Structure

According to one EPA source, the metal-finishing industry comprises more than 8,000 captive metal finishers that operate in manufacturing facilities under

TABLE 7-1. Metal-Finishing Establishments, Distribution by Size (in percent)

Size	1987	1992	1997
1–9 employees	47.1	48.7	48.1
10–49 employees	43.0	41.7	40.8
50–99 employees	6.9	6.8	7.8
100–249 employees	2.6	2.4	2.7
250 or more employees	0.4	0.4	0.4
Total	100.0	100.0	100.0

Sources: U.S. EPA 1995a, 7; U.S. Census Bureau 1999.

the umbrella of a larger company, and more than 3,000 job shops that operate mostly independently (U.S. EPA 2000, *12379–12380*). A typical job shop has 15 to 20 employees and generates $800,000 to $1 million in annual gross revenues (U.S. EPA 1995a, 5). Although employment in job shops shrunk from a total of 123,300 employees in 1988 to 74,640 in 1997, the number of job shops increased during the 1990s, from 3,294 in 1992 to 3,404 in 1997. In the same years, the value of shipments also increased, from $4.6 million to $6 million (U.S. Census Bureau 1999). As Table 7-1 shows, although there has been a slight trend over the 1987–1997 period toward increasing firm size as the number of firms has fallen, close to 90 percent of industry employees work in shops that employ fewer than 50 workers.

Metal finishing takes place primarily in industrialized states. This locational concentration occurs in part because it is cost-effective for small metal finishing facilities to be near their customers (U.S. EPA 1995a, 7). As of 1997, 70 percent of all metal-finishing businesses, accounting for over 87 percent of industry shipments, were located in eleven states: California, Connecticut, Illinois, Indiana, Massachusetts, Michigan, New York, Ohio, Pennsylvania, Texas, and Wisconsin (U.S. Census Bureau 1999). Although all of these heavily industrialized states have large numbers of metal-finishing facilities and other metal-related industries (U.S. EPA 1995a), four of them—California, Illinois, Michigan, and Ohio—dwarf all others in terms of total metal-finishing shipments, contributing almost half of the national total for 1997 (U.S. Census Bureau 1999).

Metal finishers supply a variety of industries, but the automobile industry is the metal finishing's dominant buyer, accounting for 40 percent of the metal-finishing market in 1992, followed by electronics and consumer durables (U.S. Census Bureau 1999). The metal-finishing industry depends on production by other industries, but its sales have not kept pace with sales of the industries it serves (U.S. Census Bureau 1999).

Because metal finishers tend to cluster near their industrial customers, so that competition is localized, the large total number of metal-finishing job shops may exaggerate the competitiveness of the industry. Still, based on a recent estimate that 300 shops comprise about 15 percent of the industry, it is clear that metal finishing is far from being an economically concentrated industry (Nixon 1998, *16*).

Environmental By-Products

Metal finishing is an industrial process that coats an object with layers of metal to control corrosion or friction, improve appearance, or bestow other properties. Metal finishing usually consists of three stages: surface preparation and cleaning, surface treatment, and post-treatment such as rinsing. Wastes generated through these processes include a combination of solvents and cleansers and metal-ion-bearing aqueous solutions used for plating. The most common ion-bearing metals found in solutions used for plating are hexavalent chromium, trivalent chromium, copper, gold, silver, cadmium, zinc, and nickel. Acid from the cleaners can be found in the wastewaters, and some of the solvents may be emitted into the air, released in wastewater, or disposed of in solid forms (Nixon 1998).

Typical metal-finishing wastes include wastewater treatment sludges and spent solutions, base metals, cyanide, solid wastes, and air emissions. Some processes, such as anodizing, produce acid wastes. Other processes produce alkaline wastes. Some processes produce both acid and alkaline wastes (Nixon 1998). In 1993, 2,363 fabricated metal facilities were subject to the U.S. EPA's toxic reporting requirements. These facilities released some 72 million pounds of toxic materials into the air, water, and landfills (Nixon 1998).

Federal Environmental Regulation

The incentive for firms within an industry to participate in a voluntary management-based program such as the SGP depends crucially on the default or background regulatory environment. It is therefore important to highlight key aspects of the status quo federal regulatory environment against which the SGP was formulated.

Metal finishing generates a variety of wastes and is regulated under a number of media-specific federal environmental statutes, the most significant for metal finishers being the Clean Water Act (CWA) and the Resource Conservation and Recovery Act (RCRA). This section highlights the water and waste regulations most crucial to understanding the SGP and shows how metal finishing firms have been affected by these regulations. Although the data on the effects of regulation on this industry are not extensive, their message is clear: the fear of regulatory enforcement under the status quo regulatory scheme has improved the environmental performance of the metal finishing industry, but the industry is severely limited in its ability to attain further improvements.

Regulation under the CWA. Under the federal Clean Water Act, companies that discharge wastewaters from point sources fall into two general categories: (1) direct dischargers that discharge wastewater directly into streams and lakes, and (2) indirect dischargers that discharge wastewater into municipal sewage systems. The CWA requires direct dischargers to have a permit under the National Pollution Discharge Elimination System (NPDES). These permits specify the maximum quantity of various types of pollutants that may be contained in a discharger's wastewaters. These maximum quantities are derived from technology-based effluent limitation guidelines (ELGs), which means that they represent the amount of various pollutants that may remain in a discharger's wastewater after it has been treated to remove pollutants using a specified wastewater treatment technology.

For most pollutants, ELGs determine what an economist or engineer would naturally call the "effluent standard" that is included in a particular facility's NPDES permit. Under the CWA, ELGs are categorical, with different industry categories or groups subject to different ELGs. The actual statutory language of the CWA seems to suggest that the severity of the ELGs should vary both with the type of pollutant (e.g., toxic, conventional organic, or nonconventional, nontoxic pollutants), and also with the vintage of the facility, with new facilities facing tougher "new source performance standards" (NSPS).[2] Federal courts interpreting the CWA have granted EPA vast discretion in setting ELGs for both existing and new point sources of water pollution. The legal decisions clearly establish that the agency may consider the economic impact of a proposed ELG only on an industrywide, as opposed to a plant-by-plant, basis.[3] They are equally clear in holding that the agency is under no legal duty to conduct any cost-benefit analysis at all in setting ELGs.[4] In practice, EPA has often set NSPSs at the same level as the "best achievable" or "best conventional" levels mandated for existing sources (Battle and Lipeles 1998, *172*).

This is a quick snapshot of the CWA regulatory scheme for direct dischargers. Indirect dischargers, which discharge wastewaters into municipal sewage systems for treatment by POTWs, are regulated somewhat differently. Such indirect dischargers have to comply with both federal and local standards requiring that effluents be reduced from the discharger's waste stream before it enters the POTW.[5] Like NPDES standards, these are largely technology-based. Unlike the NPDES permit program, which is implemented and enforced either by state or federal regulators (depending upon whether or not the state has a federally authorized water pollution control program), local POTWs have the primary responsibility both to implement and enforce pretreatment standards (see EPA's *Code of Federal Regulations* Title 40 [40 CFR] Section 403.8, U.S. EPA 2005). Thus POTWs are both regulated, as direct dischargers, and regulators, of indirect discharges.

Since its passage in 1972, the federal CWA has succeeded in markedly improving the quality of many of the nation's lakes and rivers. Most of this improvement came about as industrial facilities stopped discharging directly into lakes and rivers and began discharging (as the law allowed them) into municipal sewer systems. The POTWs that process these wastewaters were in turn significantly upgraded as a result of federal grants and litigation pressure.

Foreseeing an incentive for industrial wastewater dischargers to burden POTWs with their wastewater flows, in the CWA Congress instructed EPA to set pretreatment standards—the pollution-reduction standards that apply to dischargers into sewer systems—at a level equal to the NPDES standards, thereby eliminating the incentive for firms to discharge into municipal systems simply to get the benefit of a more relaxed regulatory regime. Of equal importance, however, is that federal law allows state and local water regulators to set pretreatment standards that are more demanding than the federal standards. Many of the heavily industrialized states in which the metal-finishing industry is concentrated have done precisely that.

Soon after the passage of the CWA, EPA began to write CWA regulations for the metal-finishing industry. By 1983, EPA had written two sets of regulations that remain today the most significant CWA regulations for this industry. Under these regulations, found in Part 433 of Title 40 of the *Code of Federal Regulations* (U.S. EPA 2005), EPA sets out NPDES and pretreatment standards for metal-finishing facilities. However, under the terms of a 1980 agreement settling a lawsuit brought by the National Association of Metal Finishers (NAMF) and the Institute for Interconnecting and Packaging Electronic Circuits (IIPEC) against EPA, the regulations provide that job shop electroplaters and independent printed circuit board manufacturers that indirectly discharge their wastes are exempted from Part 433. They are regulated only under rules EPA promulgated in 1979 and located in Part 413 of the *Code of Federal Regulations*, which set NPDES and pretreatment standards for existing electroplating facilities. EPA agreed to allow the existing Part 413 regulations to remain unchanged for "several years" (U.S. EPA 1980, *45323*).[6] Through the terms of this settlement, EPA indefinitely grandfathered thousands of small, indirectly discharging metal-finishing job shops on the grounds that it would be impossible for many of them to remain in business if they had to meet the Part 433 regulations that govern metal finishing more generally.

Not all firms involved in metal processing fell within the regulatory categories covered by Parts 413 and 433 of the *Code of Federal Regulations*. By 1986, EPA had formally identified a "significant number of metals processing facilities" that were discharging a "significant" amount of pollutants, and that were nonetheless not covered by either the metal-finishing industry regulations or any other related regulatory provisions (U.S. EPA 1995b, *28212*). By 1992 EPA had created a new, broader industry regulatory category called *Metal Products and Machinery* (MP&M), designed to capture many of the facilities that were currently unregulated. Having created this new category, EPA was statutorily obligated to promulgate effluent and pretreatment standards for it. When the agency proposed rules for the category in 1995, EPA retained the grandfathering of old (pre-1982) indirectly discharging job shops and independent circuit board manufacturers, and also exempted surface-finishing job shops; nonetheless, the proposed regulations would have toughened regulatory requirements for a considerable number of metal finishers (U.S. EPA 1995b, *28215*).

In 2001, EPA revised the new MP&M regulations to replace the existing regulatory scheme for metal finishers, including electroplaters. The metal-finishing industry viewed these proposed MP&M regulations for direct and indirect dischargers as a major threat. Not all metal finishers would have been equally affected by the new regulations. Indeed, EPA found that, although the technological wastewater treatment method it used to establish the new "best practicable technology" (BPT) and BCT effluent guidelines and pretreatment standards was already in use by all of the 15 directly discharging metal finishing job shops, only 55 percent of the 1,500 indirectly discharging metal finishing job shops were using it (U.S. EPA 2001a). Still, as of 2001, EPA estimated that the costs imposed by the new proposed wastewater treatment technologies would result in the closure of only 10 percent of the indirectly discharging metal finishing job shops (U.S. EPA 2001a).

By the summer of 2002, EPA was in the midst of a major reconsideration of the proposed MP&M rules (U.S. EPA 2002). In May 2003, EPA withdrew the proposed new regulations of the metal-finishing industry in their entirety (U.S. EPA 2003). What had happened to cause this dramatic reversal? Although the timing of the withdrawal—soon after Bush replaced Clinton in the White House—might suggest that political pressure was exerted within EPA, I have found no evidence to support such a suggestion. Instead, it seems that this is a case where the notice and comment rulemaking process worked as it is supposed to, giving the regulated sector a chance to amass data to persuade the agency to change course. The metal finishing trade association spent over $1 million preparing detailed economic and engineering studies showing that, because most metal-finishing job shops are engaged in a variety of different finishing operations and use a variety of different metals, it would cost them far more than EPA had assumed to meet the proposed new standard. On the benefits side, trade association consultants discovered a number of errors in EPA's estimation of the amount of wastes, such as cyanide, currently found in typical metal-finishing wastewaters. These errors tended to vastly overstate both the amount of waste currently being generated and the environmental benefits of the proposed new treatment requirements.

EPA's withdrawal of the rulemaking acknowledged that the agency had been wrong. The document explained the decision not to promulgate new metal-finishing regulations on the grounds that new data submitted in comments by the industry had caused it to (1) drastically revise upward its estimate of the number of directly discharging job shops that would close from 0 to 12 (50 percent of the reported 24 existing direct dischargers, a number that had EPA had incorrectly taken to be only 15 percent); (2) doubt the effectiveness of the proposed technology in removing metal wastes from waste streams containing a wide variety of metals; and (3) increase the number of indirectly discharging job shops that would close as a result of the regulation from 10 to 46 percent of such facilities (U.S. EPA 2003, *25704–25705*). In rejecting mandatory environmental management systems and other pollution-prevention alternatives, EPA noted significantly that "many metal finishing job shops are currently employing best management/pollution prevention practices similar to those described in the proposal as part of the National Metal Finishing Strategic Goals Program" (U.S. EPA 2003, *25692*). EPA's management-based voluntary program had effectively substituted for mandatory wastewater treatment technologies.

Regulation under RCRA. Congress passed the Resource Conservation and Recovery Act (RCRA) in 1976 with the purpose of reducing the generation of hazardous wastes and improving the way waste is handled and disposed. RCRA sets standards to govern the generation, treatment, storage, transportation, and disposal of hazardous wastes (Jalley et al. 2002).

RCRA and its legislative amendments have set up a "cradle-to-grave" system to regulate hazardous wastes. This monitoring system consists of six elements:

- identification and listing of hazardous waste,
- a manifest system to trace the life cycle of the waste from generation to disposal,
- minimum standards for waste treatment,
- storage and disposal requirements,
- requirements for state implementation of hazardous waste management, and
- an enforcement program.

The consensus view about RCRA among both academic commentators and practitioners is that, although it surely has made industries much more careful in their handling, storing, and disposing of hazardous waste,[7] the RCRA regulatory system is a "horrible" (Stoll 1989, 6) maze of "mind-numbing," complex, technology-based rules (American Mining Congress v. EPA: 1189). For the metal-finishing industry, a longstanding complaint about RCRA was EPA's rule that generators of greater than 1,000 kilograms of hazardous waste per month may not accumulate hazardous waste on site for longer than 90 days without obtaining a RCRA permit (40 CFR 262.34). The problem for generators was a simple issue of transportation costs. The roughly 1,300 metal finishers producing hazardous wastewater sludge in amounts that exceed the regulatory requirements for small-quantity generators did not generally create enough of such wastes within 90 days to fill a full truckload. To get the waste off-site without being deemed waste storage sites under RCRA, these metal finishers were shipping many partial loads. Because there is a fixed cost just to bring a truck to and from a facility, metal finishers could significantly lower their waste disposal transportation costs if they were allowed to retain waste on-site for a longer period.

In 2000, EPA revised the rule in question to allow waste generators to retain "F006 waste"—the hazardous wastes from wastewater sludge—on-site for up to 180 days, provided that they recycle the waste, implement pollution-prevention practices to reduce the toxicity of the sludge or make it more amenable to metal recovery, have no more than 20,000 kilograms of the waste on-site at any time, and comply with RCRA requirements on storage, labeling, personnel training, waste analysis, and record keeping (40 CFR 262.34(g)).

Effects of Regulation on Metal Finishers' Environmental Performance. Since the finalization of key federal regulations in the mid-1980s, the metal-finishing industry's environmental performance has improved markedly. Normalized by annual value of shipments, toxic releases from firms with emissions large enough to be reported on the federal Toxics Release Inventory (TRI) declined by 44 percent between 1993 and 2001 (U.S. EPA 2004, 33). The number of civil and criminal enforcement actions taken against firms within the industry has likewise fallen.

Federal environmental regulation seems to have led to two types of effects that are responsible for this improvement in metal-finishing industry environmental performance. The first and most direct effect was to cause the closure of many of the most marginal, worst-polluting firms that simply could not comply with federal environmental requirements. Until the 1980s, the metal finishing industry was an environmental disaster area, with typical facilities dumping and draining their wastes without treatment, and with wastes accumulating all over facility sites.[8] Many of the environmentally dirtiest firms were also economically marginal, and they closed as soon as environmental regulation became a reality.

These firms left a legacy of contaminated real estate that regulators are still dealing with today. The metal-finishing industry did not, however, disappear, and the second effect of federal regulation was to improve the environmental performance of surviving firms. Indeed, all of the metal finishers who participated in an interview study conducted by Gunningham et al. (2004) opined that their industry's improved environmental performance was mainly because of regulatory enforcement (see also Duke 1994). Indeed, almost half of their respondents said that the reason they had undertaken a specific environmental improvement was because a fine or imprisonment had occurred at their firm or at another metal-finishing firm (Gunningham et al. 2004). Reflecting the industry's relatively abysmal preregulatory environmental performance, the available evidence indicates that much of the improvement in metal finishers' environmental performance has been achieved through very simple housekeeping improvements, such as exercising more care in handling or cleanup, and basic treatment techniques, such as dewatering waste sludges (Duke 1994, *54–58*).[9]

The common complaint by environmentalists that the fines for environmental regulatory violations are too low to make a difference seems not to apply to the SMEs that make up the metal-finishing industry. A substantial number of managers of metal-finishing firms say that they personally know about instances where metal finishing firms have shut down because they were unable to pay the fines for environmental violations. Indeed, the metal-finishing industry was the only industry of six studied by Gunningham et al. (2004) in which every respondent thought that fines could cause firms to shut down. Given that fines do have bite in this sector, metal finishers tended to view the choice between regulatory compliance and noncompliance as one of whether to stay in or go out of business. For metal finishers, environmental compliance is a concern second in importance only to the general state of the economy.[10]

The Metal-Finishing SGP

The Strategic Goals Program (SGP) grew out of EPA efforts to "reinvent" its regulatory programs to improve their efficiency and effectiveness. Data on the effectiveness of the SGP as an alternative or supplement to traditional regulation are limited, but they do indicate that far fewer firms participated in the program than EPA had sought. This result is perhaps not surprising, given the economic incentives facing the metal-finishing industry and its structure.

Origin and Elements of the SGP

The metal-finishing SGP was a direct result of EPA's Common Sense Initiative (CSI), one of the early regulatory reinvention efforts launched by EPA in 1994 during the Clinton Administration (Coglianese and Allen 2004). The CSI metal-finishing industry subcommittee was composed of representatives from industry; environmental groups; labor organizations; and state, federal, and local regulators. This subcommittee developed the seven goals that became institutionalized in the SGP, which EPA created in 1998. Six of the goals were in the form of quantitative reductions relative to 1992 baseline levels:

- 50 percent reduction in water usage,
- 25 percent reduction in energy consumption,
- 50 percent reduction in land disposal of hazardous sludges and an overall reduction in sludge generation,
- 50 percent reduction in metals emissions to water and air,

- 98 percent reduction in metals use, and
- 90 percent reduction in emissions of organic toxics reported under the TRI.

The seventh, nonquantitative goal called for a reduction in human exposure to toxics both in the facility and in the surrounding community.[11] According to industry sources, the reason that the SGP backdated its baseline to 1992 (rather than using 1998, the actual year of program inception) was to avoid in effect punishing metal finishers who had already achieved significant reductions in waste generation and resource use before the program began. At the same time, although the subcommittee called upon the industry to achieve the SGP goals within 10 years, progress was to be assessed in 2002, so that the 10-year time-frame was really a 4-year period.

The subcommittee also recommended the amendment to the section of RCRA governing on-site storage of wastewater sludge and undertook a project to determine whether metal-finishing wastes had become less toxic since the rule was proposed two decades earlier. Phase I of this latter project was completed in 1998 and published as the *Metal Finishing F006 Benchmark Study* (U.S. EPA 1998). Phase II involved an effort to develop regulatory and administrative strategies to "promote metal recovery of F006 waste, encourage pollution prevention practices related to the generation of F006 waste, and reduce or remove possible RCRA barriers to metals recovery of F006 waste" (U.S. EPA 2000, *12380*). The subcommittee also identified a new chromium filtering approach that is expected to reduce chromium air emissions and cut costs by 90 percent.

The SGP's plan for accomplishing its goals consisted of two components. The first was to communicate information regarding best practices in metal-finishing source reduction and pollution prevention to as wide an audience within the industry as possible. As part of this strategy, EPA and the metal-finishing industry would work together to improve recordkeeping and tracking among industry members and recognize firms that participated in these efforts (U.S. EPA 1999). The second component of the SGP was to create a financial incentive for companies to adopt best practices voluntarily. EPA pledged to encourage POTWs that implement and enforce CWA regulations to develop a tiered approach to regulating metal finishers. POTWs would relax regulatory requirements on metal-finishing firms that had met or were on the way to meeting SGP goals. The regulations targeted for relaxation were rules that were particularly costly to comply with but not socially productive for firms attempting to go beyond compliance.

Two years into the program EPA reported that SGP had made an "impressive start," with 425 metal finishing firms, 21 states, and 80 local POTWs in the program (U.S. EPA 2001b, *5*). EPA touted the "tremendous progress" (U.S. EPA 2001b, *6*) made by industry participants in meeting the various performance goals, with the average SGP metal finisher reducing water use by 41 percent, energy use by 14 percent, hazardous sludge sent for landfill disposal by 36 percent, air and water releases of TRI metals and cyanide by 58 percent, and organic TRI releases by 77 percent (U.S. EPA 2001b). The original goal of a 98 percent reduction in the use of metals—which, given the fact that metals are the primary input to metal finishing, would seem impossible to achieve without substantial decreases in firms' production levels—was modified by the SGP steering committee to mean either a 50 percent reduction in wastewater treatment sludge, or land disposal of just 2 percent or less of the metals used by a facility (U.S. EPA 2001b, *11*). According to EPA, 112 SGP facilities met this goal by the program's midway point, primarily by recycling their sludge (U.S. EPA 2001b, *11*).

EPA's 2001 report contained several examples of how firms in the SGP achieved such large reductions in water use and waste generation (U.S. EPA 2001b, *8–12*). A firm in Ohio, for instance, was able to reduce its water usage simply by capturing all wastewater in a batch tank rather than discharging directly into the city of Cincinnati's sewers. A Colorado- and Nebraska-based nickel chrome finisher providing parts for Harley-Davidson motorcycles changed its production process by improving its monitoring of the tanks in which parts are bathed so as to prolong the life of the bath, thus reducing the amount of waste sludge created. A Portland, Oregon, firm dramatically reduced its water usage by installing two cooling towers. Other stories involved companies that had reduced the amount of wastewater sludge sent to landfills by drying the sludge and sending the dried sludge to metal-recycling facilities.

Perspectives on the SGP's Performance

Although EPA's 2001 report on the midterm progress of the SGP touted the program's success, a closer look in light of more recent data reveals a more complex story. Data current to 2003 show, for example, that although the goal of a 50 percent reduction in metals emissions to air and water was exceeded (with a 57.8 percent reduction by 2002), the program fell short of its goal of a 50 per cent reduction in sludge generation and land disposal of sludges, with sludges reduced by only 3 percent and land disposal of sludges by 15 percent.[12] The program's participation rate, however, was far short of its goal of roughly 1,050 (80 percent of a conservatively estimated total number of 1,300).[13] Indeed, although EPA's *Living the Vision* 2001 progress report on the SGP claimed that 425 metal finishers were participating in the program (EPA 2001b, 5), its national dataset shows that the number of metal finishers who were actively and fully participating in the program—by turning in annual reports—reached a peak of 226 in 2000, with only about 113 still reporting at its conclusion in 2002.[14] Given the likelihood that the most-committed, best-performing firms were the ones that took the time to fill out annual reports, EPA's data likely overstate the program's achievements.

In order to fully evaluate both the SGP's relative success in stimulating waste reduction and its apparent failure in getting high participation, one needs two things: more detail on the numbers, and benchmarks for comparison. An initial problem with the waste reduction numbers is that EPA never verified them. According to the industry's trade association, the methodology for generating and checking the numbers was produced by an outside consultant. Because participation in the SGP was voluntary, reliance upon an outside contractor is perhaps the best that could be hoped for, and not materially different from what EPA would have done on its own.

A bigger problem for ex post assessment of the SGP is its failure to generate firm-specific data. Ideally, one would have detailed firm-level data not only on the level of wastes generated both before and after the initiation of the program, but also on other firm attributes that theory suggests should be relevant to a firm's decision about whether to participate. Similarly, detailed firm-specific data on firms that did not participate in the SGP would then permit econometric testing of various theoretical hypotheses about why firms participated as well as testing whether participation had a statistically significant effect on waste reduction performance over the relevant time period.

The primary reason why the SGP did not generate such firm-specific data is that firms agreed to participate only on the condition that such data would not be made publicly available. One can understand why metal finishers would have preferred such nondisclosure: environmental groups and others who had never "signed on" to the SGP could use SGP reports in citizen suit litigation. But EPA's agreement to such nondisclosure is questionable. Completely aside from legal and ethical issues—can, for example, EPA evade Freedom of Information Act disclosure obligations by putting data collection and dissemination in the hands of a separate entity such as SGP?—the agency cannot learn whether voluntary programs such as the SGP work or not unless such programs generate firm-specific data. Moreover, it may be possible to provide such information without revealing firm identity simply by assigning each firm a coded identity.

In the absence of firm-specific data from the SGP itself, one is left to draw what inferences one can from the data that SGP did provide, and from data about metal-finishing environmental performance generated independently of the SGP. In evaluating the program's aggregate performance, a useful benchmark is provided by a study done in 2000 by the National Center for Manufacturing Sciences (NCMS 2000). That study produced metal-waste reduction goals for zinc, decorative chromium, electroless nickel, and hard chromium plating. These goals were set equal to the performance of the top-performing quartile of metal-finishing firms in a sample of such firms conducted by the NCMS. Under that methodology, the NCMS study recommended sectorwide reduction goals of 54 percent for zinc wastes, 73 percent for nickel wastes, and 71 percent for chromium wastes. Recall that the SGP set a goal of 50 percent reduction in metal wastes and reduction of simply any magnitude in sludge. The SGP's data claim a 16 percent reduction in sludge and 37 percent decrease in wastewater. Although the sludge reduction meets the SGP goal—which was any reduction at all—according to the SGP data, the wastewater reduction falls below the SGP goal. More importantly, the SGP's goals were much more modest than the waste-reduction objectives recommended by the NCMS study: 50 percent versus an NCMS average of 65 percent. Compared with the NCMS objectives, the SGP appears to have failed to set its sights high enough in the first place, and then to have fallen short of what the NCMS viewed as achievable by the typical firm within the industry.

The only firm-specific data I have found comes from a pollution-prevention report issued by the state of Minnesota's Office of Environmental Assistance in 1999 (MOEA 1999). This report presents TRI data for the 26 Minnesota metal-finishing firms (both captive and job shop) that reported TRI data in 1999. Six of these firms participated in the SGP. I compared the environmental releases of SGP and non-SGP firms included in this study. The sample is small, and I did not identify which firms on the Minnesota list were captives and which were job shops. With these caveats, Figure 7-1 compares the average change in TRI emissions between 1995 and 1999 for SGP and non-SGP metal finishers in Minnesota.

The data reported in Figure 7-1 suggest that for cyanide and nitric acid emissions, Minnesota SGP participants were doing better than non-SGP metal finishers, although they had a worse record in reducing nickel emissions. This conclusion depends, however, on excluding from the sample one SGP firm that increased its cyanide emissions by 1,220 percent over the period. If that firm is included, then SGP firms in Minnesota did better than non-SGP firms only for nitric acid.

It may be argued that Figure 7-1 is not relevant to evaluating the SGP because it does not cover the entire period of the program. However, the SGP's own reported aggregate data raise questions about whether the SGP actually helped firms improve their performance or simply attracted firms that were already good performers. As observed by Coglianese and Allen (2004), these data show that most SGP firms made little progress in achieving program goals after the SGP was launched in 1998, with much of the progress that did result occurring before the program even began. For instance, normalized (by sales) hazardous waste sludge generation by SGP firms went down by 16 percent in the first two years of the program, but in the program's third year this went back up to its initial level (Strategic Goals Program 2001b). Similarly normalized toxic organic chemical emissions went down until 2000, but then increased by 30 percent (Strategic Goals Program 2001c). Only water discharge, which went down 37 percent relative to 1992 baseline usage, fell continuously during the entire life of the SGP period (Strategic Goals Program 2001d). Whether this one unambiguous change resulted from the program, however, is unclear: by 1996, over 90 percent of metal-finishing shops had already introduced measures to

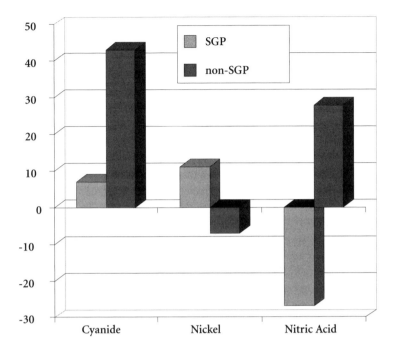

FIGURE 7-1. Minnesota Metal Finishers 1995–1999 Change in Selected TRI Emissions
Source: Adapted from MOEA 1999.

reduce dragout and rinse water volumes (Washington State Department of Ecology 1996).[15]

As to the question of why some firms joined and some did not, there is once again a dearth of systematic evidence. Although many firms in Indiana and Michigan participated, few firms in other states with large metal-finishing industries chose to take part. Of the 23 metal finishing job shops in the state of Washington as of 1999, only 2 were SGP participants. In California, the state with the largest metal-finishing industry, only 12 shops participated in SGP. Most of the 16 biggest generators of hexavalent chromium and cyanide wastes in the California metal-finishing sector are captive metal finishers (e.g., Boeing Satellite Center), but several are independent job shops, and none participated in the SGP (CA DTSC No date). At the conclusion of a multimedia enforcement effort in 2000, the "general perception" of California environmental regulators was that the chromium-plating industry in that state "ha[d] compliance problems in all environmental media" (CA DTSC No date, 5). The general pattern seems to be one in which participation in the SGP clustered in certain states and localities.

Explaining the SGP: An Economic Approach

The field of economics offers explanations as to why individual firms and their trade associations might participate in voluntary management-based strategies such as the metal-finishing SPG. In this section, I review these explanations and explore the extent to which an economically motivated account can explain some of the general features of this program.

Economic Incentives for Management-Based Voluntary Programs. The SGP was one of many "voluntary" environmental programs initiated during the Clinton Administration. As described above, it emerged from that administration's Common Sense Initiative (CSI). Like other voluntary programs of that time period, such as EPA's Project XL, the SGP involved a commitment by participating metal-finishing firms to improve their environmental performance beyond what was legally required.[16] Unlike Project XL, however, which involved negotiated agreements between federal, state, and local regulators and individual facilities, the SGP was a sectoral initiative. Explaining the SGP requires understanding both the general incentives for individual firms to go beyond existing regulatory requirements and the particular incentive issues raised by attempts to improve the environmental performance of an entire industry sector.

Firms may undertake any number of costly activities, for any number of reasons. If a firm has substantial market power in one or more industries and is generally well run, it will typically have plenty of free cash flow. This cash flow gives the senior management of such companies discretion over company policy, particularly closely held firms that are not subject to the discipline imposed by public stock markets. For publicly traded firms, the stock market may eventually discipline policies that harm the bottom line by lowering a company's stock price, but the stock market's ability to discipline inefficient management is contingent upon a number of variables. Most important among these variables is the complexity of the firm's financial structure and transparency of the firm's financial performance. Given the both substantial discretion that senior corporate managers enjoy over policy choice and their hierarchical control position in their organizations, a firm may voluntarily improve its environmental performance simply because its current CEO has decided to make the environment a priority regardless of how such a priority affects the firm's bottom-line profitability (Kagan et al. 2003).

The existing economic literature on voluntary environmental programs has devoted relatively little attention to managers' altruistic motives for participation in such programs. The literature has instead discussed a variety of self-interested reasons why individual firms might participate in voluntary environmental programs. There is, on the one hand, the *green gold hypothesis* associated with Michael Porter (Porter 1991; Porter and van der Linde 1995), which holds that by trying to improve their environmental performance, firms often discover previously unknown ways of lowering their costs and increasing their productivity. Although case studies suggest that such opportunities have sometimes been realized, economists by and large remain skeptical that merely by trying to improve their environmental performance, firms can discover opportunities to lower costs (Palmer et al. 1995; Boyd 1998).

By contrast, economists have shown that the presence of consumers and investors who care about firm-specific environmental performance, and who will pay a premium price for the products or stock of firms with superior environmental performance, may create a direct profit motive for a firm to establish a reputation as a strong environmental performer (for a rigorous demonstration of conditions under which this motive exists, see Small and Zivin 2002). However, socially responsible consumers and investors may not create an incentive for actual investment in improved environmental performance, but instead only for investments in public relations to create a widespread perception of such improved performance that may belie a quite different reality (Johnston 2005a).

Regardless of the reality behind such perceptions, there is one important type of customer demand for such improvements: large firms that buy parts, products, and materials from smaller suppliers. Originating in the goal of quality control, *supply chain management* has become a tool by which large firms attempt to improve the environmental performance of their typically smaller suppliers. There are a number of reasons why such big firms might want to push their suppliers to improve their environmental performance, ranging from concern that laws may someday change to hold them liable for the environmental harms caused by their suppliers (a specter made quite thinkable by the Superfund law's regime of strict, retroactive, and joint and several liability for hazardous waste cleanup) to more immediate concern about their own image with consumers and investors. Whatever the reason, if big customers demand environmental performance as a condition for continuing to buy, small suppliers with serious competitors have an obvious incentive to accede to such demands.

A third economic reason that firms participate in voluntary environmental programs is that they view such programs as a strategic tool, to be used like more traditional lobbying and litigation to persuade politicians and regulators to relax or at least not toughen the firm's environmental regulatory compliance requirements. On this explanation, participation in voluntary environmental programs is a form of *strategic regulatory preemption* (Segerson and Miceli 1998; Maxwell et al. 2000). The difficulties that regulatory beneficiaries face in securing new, tougher regulatory standards means that firms may readily stave off new regulation by voluntarily committing to levels of performance that are higher than those currently required but lower than those that regulatory beneficiaries prefer.

All three economic explanations help to illuminate the behavior of a manager of an individual firm faced with deciding whether to join a voluntary environmental program. However, two of these economic explanations are not likely to operate at the level of an industry sector such as metal finishing. In environmental regulation, *industry sectors* are collections of firms that may or may not be direct product market competitors, but they use sufficiently similar production processes to have to comply with the same set of technology-based regulatory standards. Industry sectors take action through their trade associations (Johnston 2005b). If firms are competing against one another, they have an incentive to find new and better ways to lower costs or increase revenues, and they will treat those ways as proprietary information not to be shared with competitors. A firm will not want an industry trade association to share with its competitors information about cost savings it obtains through pollution prevention or any revenue increases it reaps by attracting green consumers. In other words, while green gold and green consumers and investors may motivate individual firms to participate in voluntary environmental programs, these reasons are unlikely to account for industry trade association involvement in or sponsorship of such programs.

These economic reasons for complying with voluntary programs do not apply to strategic regulatory preemption. Under the existing regulatory regime, firms within a given regulatory sector face the same technology-based regulatory standards. Proposals to toughen regulations threaten all firms within a sector with a regulatory cost increase. All firms within the affected sector share a common interest in contributing to a voluntary environmental program that might preempt such costly proposals from being finalized.

These basic theoretical predictions about individual firms and their trade associations follow only under the assumption that the sector consists of identical firms that are product market competitors. Under a different set of assumptions, dominant firms within trade associations may have an incentive to use the association to publicize the cost-saving and revenue-enhancing advantages of improved environmental performance among association members. Most simply, if firms within the association are not direct product market competitors, but simply fall into the same regulatory category, then they will suffer no competitive loss by sharing information about how to achieve better environmental performance. Even if firms do compete in product markets, larger firms within a trade association that tend to have high public and regulatory visibility may have a distinct incentive to use the trade association to promote compliance with voluntary environmental programs by smaller firms within the industry. As observed by King and Lenox (2000), a large firm such as DuPont may bear a disproportionate share of the collective reputation of firms within the chemical industry, and for this reason such a large firm may have an incentive to take steps through the industry trade association to improve industrywide environmental performance.

Trade association activity may also be both motivated and justified by the public good aspect of investments in pollution prevention. By patenting its invention, a firm that discovers a new pollution abatement technology can capture close to the full value of its invention by selling it to other firms that wish to adopt it. Of course, new production methods or management practices that prevent pollution or reduce raw materials usage are generally difficult to patent. A firm that discovers such new methods may have difficulty preventing other firms from simply copying what it is doing. If pollution prevention exhibits the characteristic of nonexcludability, then pollution-prevention innovators will have an inadequate incentive to innovate. Nonexcludability might also imply that there is little a trade association can do to overcome this underinvestment problem.

The conclusion that there is little a trade association can do to encourage its members to prevent pollution neglects the information role played by such associations. Although some trade association members may find out quickly on their own about new ways to prevent pollution and cut materials usage, others may never get such information unless their trade association provides it to them. For many members, investments in an effective trade association are a way to get information that can directly improve their economic performance. The financial contributions of its members may thus allow a trade association to subsidize the creation and dissemination of information about new pollution-prevention methods. Although the evidence is at best mixed on whether voluntary environmental programs initiated by trade associations lead to real improvements in the environmental performance of participating firms (Howard et al. 2000), there are instances in which trade associations have acted to publicize and promote the adoption of improved environmental practices within their industry, even if these might improve the competitive position of some members (Schwartz et al. 1999).

Economic Reasons for the Metal-Finishing SGP. Although I have been unable to discover any systematic study of the economic structure of the metal-finishing industry, anecdotal data suggest that because of the relatively high cost of transporting metal products to and from finishing shops, such shops will be located close to their customers, so that competition within the industry may be expected to be highly localized. Anecdotal data also suggest that the metal-finishing industry may consist of two quite different types of firms: a lower tier of firms that provide a relatively low-quality, low-price product and that have little competitive slack, and an upper tier that charges premium prices for high-quality products.

There has, by contrast, been more systematic study of what may be called the regulatory structure of the metal-finishing industry. By *regulatory structure*, I mean the way in which metal-finishing firms differ in their attitude toward environmental regulation and their history of compliance. The metal-finishing industry was one of the first industries studied as part of EPA's Sustainable Industry Project, and the agency's report on metal finishing portrays distinct tiers of firms within the industry (U.S. EPA 1994, *25–28.* These tiers bear a striking resemblance to those Kagan reports on in his study of pulp and paper mills in Chapter 2 of this book. At the highest tier are firms that have been regulated and expect to continue to be regulated in the future, and that also expect a long future in business. Although managers of these firms may differ in how they express their environmental attitudes—some say they care about their environmental performance simply because it is the "right thing" to do, while others express more self-interested motives—they all share a strong interest in maintaining a cooperative, long-term relationship with regulatory authorities. Because they anticipate a continuing regulatory relationship, such firms are willing to go beyond what regulators may currently require.

At the next tier are firms that will comply with regulations only out of fear of enforcement. The ability of these firms to comply (let alone go beyond compliance) may be severely limited because of their relatively precarious financial state. When it comes to the sorts of SMEs that make up the metal-finishing industry, these firms—what Kagan in Chapter 2 calls the "reluctant compliers"—are more the rule than the exception.

The final and lowest tier of firms is made up of those firms that simply do not comply with environmental regulations. They get away with noncompliance because regulators with limited budgets simply cannot inspect and enforce against every firm that is out of compliance, or because the firm actively takes steps to avoid detection by regulators. For these noncompliers, avoiding regulatory compliance costs may be a key aspect of an overall competitive strategy to undercut competitors' prices.

There is a parallel between my conjectures regarding the economic structure of the metal-finishing industry and the evidence on the existence of tiers in the industry's regulatory structure. This parallel generates the following predictions about the metal-finishing SGP:

- Metal finishing markets are both local and specialized. The composition of firms within any given local market is therefore likely to vary with the nature of the customer base in that market. Within any geographic market, the greater the environmental sensitivity of a metal finisher's customers, the stronger the incentive for a finisher to participate in the SGP. In some markets, certain job shops may do such a high percentage of their work for one or two large customers that they are de facto captive. For such shops, supply chain pressures may strongly influence their incentives for environmental performance.

- Although supply chain incentives are customer-driven, these incentives for participation are likely to vary with location and a particular shop's customer base. Both the green gold and strategic regulatory preemption incentive for participation are likely to be relatively constant across geographic areas.

- Within any given metal-finishing market, the long-term, relatively high-quality firms that are also top-tier regulatory performers would be the most likely to participate in the SGP. Not only would such a program provide them with a relatively low-cost way to signal their environmental commitment to local POTWs, but it would also provide them with an opportunity to participate in strategic regulatory preemption with respect to federal environmental regulators. Moreover, because such firms have shouldered the burden of federal and local regulators' perceptions about the poor environmental performance of the metal-finishing industry, they would have had an incentive to support any program, such as SGP, that promised to improve the performance of lower-tier firms.
- The next tier of firms—those with lower quality but also lower price—face great economic uncertainty and therefore have a relatively short horizon in thinking about investing in potentially costly voluntary programs such as the SGP. Such firms would generally be expected to free ride on the lobbying efforts of the industry trade associations. The SGP held out the potential of benefiting firms by persuading local POTWs to adopt a more flexible regulatory approach. It also offered financial and technical assistance in pollution prevention. Indeed, the dominant reason for such second-tier firms to participate would probably have been the SGP's potential to provide green gold in the form of information about money-saving pollution-prevention techniques.
- For the bottom tier of noncomplying firms, the SGP would have been irrelevant. Since the strategy of such firms is to evade regulation, participation in a voluntary regulatory program would be nonsensical.

Although data limitations make it impossible to test these conjectures systematically, it is possible to test them indirectly by comparing them with the stylized account of the metal-finishing industry's response to, and shaping of, the SGP. The regulatory regime against which metal finishers ought rationally to have evaluated participation in the SGP was one of extremely high regulatory visibility and the specter of even greater regulatory stringency as EPA moved toward implementation of effluent and pretreatment CWA regulations for captive and job shop finishers. The simplest case in which profit-maximizing firms will agree "voluntarily" to go beyond existing environmental regulatory requirements is when new regulations would soon force them to improve anyway. In this case, "voluntary" improvement does not involve any marginal cost relative to the cost that the regulations would soon require, but brings the marginal public relations benefit of allowing the firm to tout its record in going "beyond what the law requires." Given that EPA had in fact proposed new, tougher regulations in 1995, it is likely that some metal finishers would have agreed to participate in the SGP simply to receive good publicity from doing what they expected regulations would soon require.

The regulatory history discussed above suggests that strategic regulatory preemption was a significant motive behind the SGP, both from a sectoral point of view and from the point of view of particular, larger firms in the industry. Throughout EPA's consideration of whether to promulgate new effluent and pretreatment standards for metal finishers, the industry trade association and individual firms argued repeatedly that the industry should be given a chance to show that it would voluntarily improve its environmental performance through the source-reduction and pollution-prevention steps promoted by the SGP, and that new regulations were simply not needed. EPA's May 2003 decision not to revise effluent and pretreatment standards for the metal-finishing industry explicitly noted the improvements the industry was making on its own under the SGP. Finally, although it may be coincidental, the SGP was terminated at precisely the time when EPA decided not to promulgate the new, tougher effluent and pretreatment standards.

That self-interest in forestalling regulation may have been part of the motive for the metal-finishing SGP does not mean that the source reduction and pollution prevention undertaken pursuant to the SGP was a facade. One way to view the metal-finishing SGP is as a deal between federal EPA and the metal-finishing industry under which EPA agreed to refrain from toughening traditional regulatory requirements in exchange for the industry's promise to pursue source reduction and pollution prevention. By generating improved environmental performance while dispensing with many of the costs of a permit-based system (such as sampling, inspections, and reporting), such a sectoral environmental contract can generate better environmental performance while greatly lowering administrative costs. To the extent that the SGP represented such a win-win situation, the self-interest of the metal finishers in forestalling costly new regulation matched the public interest in reducing the industry's environmental impact at lowest cost.

The most striking example of such a sectorwide agreement generated by the SGP was EPA's RCRA rulemaking regarding the storage of F006 hazardous wastes. EPA's decision to extend the period of time that metal finishers may store F006 waste on-site in exchange for pollution prevention and waste recycling efforts by metal finishers represents a classic instance of the kind of win-win situation that the inefficiency of existing regulation makes possible. In this case, the inefficiency of the status quo regulatory regime, specifically the way in which it forced metal finishers to incur unnecessary waste disposal transportation costs, was the pathway to creating financial incentives for improved environmental performance by metal finishers. Under the new regulation, more liberal on-site waste-storage time limits benefit waste finishers' bottom lines, while their commitment to recycling and waste reduction provides the environmental outcomes that are the ultimate goal of RCRA.

The SGP also offered the potential for metal-finishing firms in the top two regulatory tiers to signal their commitment to environmental performance to local POTWs in order to continue dealing cooperatively with such regulators. A core aspect of the SGP was the involvement of local POTWs in crafting a new and more flexible regulatory environment for firms that met (or began to meet) SGP goals. "Regulatory flexibility" in the SGP meant that state and local regulators would tailor regulatory requirements and enforcement penalties to a particular facility's performance. Under this approach, which may be called "regulatory tiering" (Speir 2001), facilities would be rewarded for participating in the SGP and meeting (or making progress toward) the SGP source- and waste-reduction goals. Like Project XL and other EPA regulatory reinvention programs of the 1990s, the underlying idea was to make the inefficiency of the existing regulatory structure into a virtue. State or local regulators could agree to reduce some costly but environmentally unproductive regulatory requirements (involving such things as recordkeeping, monitoring for pollutants that were not actually present in wastestreams, and unproductively frequent self-sampling) for those facilities that made progress toward meeting SGP's goals of improved environmental management and performance.

States and municipalities that appear to have been the most active in the SGP program did create regulatory tiers for metal finishers under which the better a company's performance in terms of pollution prevention and reduction, the greater the benefits it would receive.[17] These benefits were in the form of both technical and financial assistance and regulatory relaxation. Potential regulatory benefits to metal finishers were to be granted by the local POTWs who effectively regulate metal finishers as indirect dischargers. These regulatory benefits consisted primarily of reductions in the required frequency of self-sampling and inspections, with extended and simplified permits offered to metal finishers that achieved the highest level of performance.[18] In New York City, for example, self-sampling requirements for jewelers were reduced to the federal minimums, as were POTW inspection and monitoring requirements (Heckler and LaGrotta 2000).

Under the SGP, as under other regulatory reinvention initiatives, facilities were not to get regulatory benefits unless they first demonstrated progress toward or achievement of superior environmental performance. Such progress requires a facility or firm to possess both information about potential performance improvements and the capital to make those improvements a reality. As with other sector-based voluntary environmental programs, the metal-finishing trade associations cooperated with the federal EPA to produce and distribute information regarding best environmental practices. These trade associations included the American Electroplaters and Surface Finishers, the Metal Finishing Suppliers' Association, and the National Association of Metal Finishers. Through efforts ranging from pollution-prevention workshops to free nonenforcement audits, the SGP facilitated efforts by EPA and metal-finishing trade associations to produce and widely disseminate information on best environmental practices.

From the economic point of view of both the industry and society at large, there is indeed a strong case for providing such information as a free public good. From the point of view of the industry, the more widely best-practice information is shared and adopted, the better the industry's overall environmental performance, and the lower the probability of costly enforcement actions and new regulations. Moreover, although all firms within a sector benefit to some extent when the sector improves its environmental performance, the biggest firms that are most closely watched by regulators and citizen enforcers and that have the financial capability both to pay fines and meet new, tougher standards benefit the most.

Interviews with metal-finishing industry representatives confirm that it was the higher-quality, top-tier metal finishers that were pivotal in creating the SGP.[19] The evidence also shows, however, that the financial and technical assistance offered by the SGP was insufficient to enlist large numbers of the important middle-tier firms. The SGP may well have informed many such metal finishers about new and better production and waste management processes, but to succeed on a large scale, the program would have needed to subsidize substantially the investments such changes require.

Indeed, perhaps one of the clearest lessons of the SGP is that regulatory incentives alone will typically be insufficient to motivate firms to "go beyond compliance." This important point is brought home best by a concrete example. According to local POTW officials interviewed for this study, "compliance" with CWA effluent standards already requires metal finishers to generate very clean wastewater (as one source put it, it must be "very close to meeting drinking water standards"). Further reductions in wastewater effluent (to make the wastewater possibly even suitable for recycling) typically cost on the order of $75,000 to $100,000 in capital costs. By comparison, even if a POTW reduced the frequency of metal finishers' required self-sampling to federal minimum requirements, it would save a metal finisher roughly on the order of $1,000 to $5,000 per year (depending upon the status quo sampling frequency).[20] Given an industry that is already highly regulated, as is the metal-finishing industry, reducing pollution "beyond compliance" has high marginal cost. In the metal-finishing industry, at least it seems that this cost generally dwarfed the monetary value of marginal regulatory benefits that state and local regulators were able to offer by reducing regulatory paperwork, sampling, monitoring, and inspection burdens.

Should EPA Attempt to Replicate the SGP? To think sensibly about whether to try to extend a program such as the metal-finishing SGP to other sectors, EPA must answer a question that apparently no one connected with the program has ever openly asked. This question is whether it is better to try to help an industry of small- and medium-sized firms overcome its environmental management problems or instead to adopt a regulatory attitude that creates incentives for the consolidation and vertical integration of that industry into firms more capable of managing their environmental impacts. EPA targeted the SGP at the small, independent job shops, not at the metal-finishing shops that are owned by and located within the plants of large manufacturing firms such as Boeing and General Motors. Although EPA puts metal finishing that takes place within such a captive environment in a completely different regulatory category than job shop finishing, from the point of view of the environment and human health, what matters is the industrial process, not its location. Metal finishing is a process with a very high potential environmental impact. American society will continue to demand declines in this impact. As a number of empirical studies have shown (Millimet 2003), pollution-prevention and control expenditures increase the minimum efficient scale of facilities. As the experience of the metal-finishing SGP confirms, the burden of such costs is often too much for small firms to be able to finance. The natural solution, and the one provided by the market, is for such small firms to disappear and for the large manufacturing firms that have substantial internal and external sources of capital to bear the cost of society's continuing demand for improvement in environmental performance.

Regulatory policies that actively encourage firms to integrate vertically and take direct control over environmentally sensitive aspects of their production processes may be relatively rare, but those that do exist are certainly well known. For example, whatever its problems, Superfund's liability regime has completely changed the incentives faced by large chemical and petrochemical companies. Such firms no longer contract out their waste disposal operations to small, undercapitalized disposal firms, but exercise direct control over waste disposal. A similar approach—in this case involving upstream liability for the environmental harms caused by growing animals in warehoused pens—seems the likely solution to the problem of regulating concentrated animal feed operations (CAFOs). Indeed, a very basic economic lesson is that so long as large firms can effectively escape responsibility for the environmental compliance burden associated with a particular stage of their production or distribution by contracting out that stage to smaller outside firms, they will have an incentive to do so. But when the small firms cannot meet those burdens while remaining economically viable, such contracting out effectively eliminates compliance. Small economically stressed firms will usually opt out.

Most federal environmental statutes do not create a broader liability scheme. In particular, the CWA, arguably the most significant environmental law for metal finishing, does not utilize liability as a strategy. However, even under this statute, EPA retains substantial discretion. In my view, EPA should exercise that discretion not with a goal of perpetuating inefficiently small and environmentally constrained independent metal-finishing job shops, but with a goal of creating incentives for large manufacturers that now contract their operations out to such firms to internalize those operations and manage them more responsibly. The goal should be to facilitate this economically inevitable and environmentally sound transition by easing the financial costs of such transition to the small job shops.

A model for how to do this already exists: the tradeable quota market that EPA used with such great success in accomplishing the phase-down and eventual elimination of lead additives in gasoline. As described in considerable detail by others (Nichols 1997; Hahn and Hester 1989), by allowing small refiners to buy lead additive permits from large refiners that had already reduced lead content, EPA's lead phase-down rule allowed small refiners that could not financially comply with the lead additive reduction a period of years over which to continue in business profitably. Because firms that surpassed the goals could sell their unused quotas, EPA's rule also speeded up the achievement of the overall goal of eliminating lead.

Such programs are not always the right way to go. In particular, they are not appropriate when pollution is effectively concentrated in particular "hot spots." Still, they hold enormous potential for easing the transition to new, tougher environmental requirements and a new, vertically integrated metal-finishing industry. In my view, if Congress and the American public really want a dramatically cleaner metal-finishing industry—an industry that, for example, uniformly engages in costly recycling of its wastewaters—then instead of creating voluntary programs such as the SGP, EPA's time would be better spent exploring the feasibility of such a tradable rights regime for metal finishing. Rather than developing more voluntary programs, the best strategy to improve environmental management in sectors like metal finishing may be to impose mandatory performance requirements but give firms the flexibility to meet them by buying permits on the open market.

Acknowledgments

Kalpana Kotagal, Penn Law '05, provided exceptionally helpful research assistance on this project. I am grateful to Jay Benforado and Christian Richter for helpful and insightful comments made at the conference where the initial draft of this chapter was presented, to Jeffrey Smith for comments on an earlier draft, and to the editors for their extensive and careful work on this entire project. I am also grateful to the large number of EPA, state environmental and local POTW officials who took the time to talk to me about the Strategic Goals Program. I am, of course, solely responsible for any remaining errors.

Notes

1. It is important to note that under EPA regulations found at 40 CFR Section 403.8, only those POTWs with a total design flow capacity of more than 5 million gallons per day must have an approved pretreatment program for the regulation of discharges into their system by industrial users such as metal finishers. Although EPA estimates that the vast majority of indirect industrial discharges by volume are indeed subject to POTW pretreatment regulation, some environmental groups have argued that in certain states, such as California, most indirect discharges are not in fact regulated or monitored under the Clean Water Act. See Battle and Lipeles 1998, *372*.

2. For instance, according to the statutory language, existing point sources of conventional pollutants must meet effluent limitations determined by reference to the "best conventional technology" (BCT), where this is to be determined by looking at, among other things, "the reasonableness of the relationship between the costs of attaining a reduction in effluents and the effluent reduction benefits derived, and a comparison of the cost and level of reduction of discharge of such pollutants from the discharge of publicly owned treatment works to the cost and level of reduction of such pollutants from a class or category of industrial sources, and shall take into account the age of equipment and facilities involved, the process employed, the engineering aspects of the application of various types of control techniques . . ." 33 U.S.C. Section 1314(b)(4)(B). Any point source constructed after the promulgation of such a BCT-based ELG is a new source, and must, according to the statute, meet a "standard for the control of the discharge of pollutants which reflects the greatest degree of effluent reduction which the Administrator determines to be achievable through application of the best demonstrated technology" 33 U.S.C. Section 1316(a)(1).

3. *E.I. du Pont de Nemours v. Train*, 430 U.S. 112 (1977).

4. *CPC Intl. Inc. v. Train*, 540 F.2d 1329, cert. denied, 430 U.S. 966 (1977).

5. For a succinct description of the three types of pretreatment requirements, see Battle and Lipeles 1998, *356*.

6. In justifying the terms of the settlement, EPA explained its reasoning as follows:

"EPA is sensitive to the fact that the job shop metal finishing segment is vulnerable to adverse economic impacts as a result of pretreatment regulations. In the preamble to the September 7, 1979, standards, EPA estimated that 587 metal finishing job shops, employing 9,653 workers, may close as a result of these regulations. As to this segment of the metal finishing industry that is economically vulnerable, EPA does not believe that more stringent regulations are now economically achievable. Therefore, EPA does not plan to develop more stringent new pretreatment standards for the job shop metal finishing segment in the next several years. Nor does EPA plan to develop in the next several years more stringent standards for the independent printed circuit board segment, where significant economic vulnerability also exists" (U.S. EPA 1980, *45323*).

7. In the case of metal finishing, Duke (1994, *59*) concludes his study of metal finishers' waste-minimization practices by remarking that "economic incentives to avoid costly and burdensome RCRA regulations do encourage hazardous waste generators to minimize their waste streams."

8. As one shop owner put it, "17 years ago . . . we were accumulating waste under the decks and there was two feet of sludge under the floor from drips and spillage." Another commented: "If you had to get rid of something you just dumped it" (Gunningham et al. 2004, 7).

9. On the other hand, Duke et al. (1998, 36) found that over the period 1993–1995, stormwater pollution loads increased in a sample of 130 Los Angeles metal finishers and fabricators. The stormwater regulatory regime, however, is far different in several respects from the regimes governing air and water pollution; moreover, Duke et al. did not attempt to control for a number of potentially confounding factors.

10. See Finisher's Management, State of the Industry: 2001 Survey on Environmental Compliance, http://www.finishers-management.com/novdec2001/sofi2002.htm (accessed December 1, 2003).

11. For a statement of these goals, see http://www.strategicgoals.org/coregoals.cfm (last accessed February 1, 2005).

12. All these goals are in terms of sales normalized figures. The most recent data on metals discharged to air and water, sludge generation, and sludge disposal are available at http://www.strategicgoals.org/reports2/t9.cfm?state=all&requesttimeout=300; http://www.strategicgoals.org/reports2/t5.cfm?state=all&timeout=300; http://www.strategicgoals.org/reports2/t6.cfm?state=all&timeout=300 (last accessed February 2, 2005).

13. The number of 1,300 metal finishing job shops is taken from EPA's May 13, 2002, Final Rule announcing its decision not to revise CWA standards for the metal-finishing job shop industry (U.S. EPA 2003). This number differs from the number of facilities included in the 3471 Standard Industrial Classification (SIC) category that is reported in the Census of Manufacturers, because the census number includes captive shops (U.S. Census Bureau 1999).

14. The number of participants is constant for all reports except energy use, where even fewer firms reported. For the representative numbers, see, for example, the water discharge trend report, available at http://www.strategicgoals.org/reports2/t4.cfm?state=all&timeout=300).

15. *Dragout* refers to the contamination of rinse water by plating-process solutions that drip from parts being plated. Reducing dragout reduces the amount of chemicals and water used in the production process.

16. I have taken this very useful general definition of a voluntary approach from Carlo Carraro and Francois Leveque (1999). In the taxonomy provided by Thomas P. Lyon and John W. Maxwell (2004, 4–5), the SGP would be classified as a public voluntary agreement.

17. For examples, see Michigan Metal Finishing Strategic Goals Program 2000, 3–4; City of Statesville, N.C. Cooperative Agreement, 3; and New York State Strategic Goals Program 2000, 7.

18. See, for example, Michigan Metal Finishing Strategic Goals Program 2000, 4; National Metal Finishing Strategic Goals Program 2001, 4; and New York State Strategic Goals Program 2000, 7.

19. The finding that it is larger firms that tend to participate in voluntary programs is

quite general. Videras and Alberini (2000) found that larger firms were significantly more likely to participate in three different EPA voluntary programs: WasteWise, Green Lights, and 33/50.

20. See also Heckler and LaGrotta 2000, reporting that reducing the self-sampling frequency for New York city jewelers saved them an average of $1,000 per year.

References

Battle, Jackson B., and Maxine L. Lipeles. 1998. *Water Pollution.* 3rd edition. Cincinnati, OH: Anderson Publishing.

Boyd, James. 1998. *Searching for the Profit in Pollution Prevention: Case Studies in the Corporate Evaluation of Environmental Opportunities.* Washington, DC: Resources for the Future.

CA DTSC (California Department of Toxic Substances Control). No date. Metal Plating and Finishing Industry Project Feasibility Overview. http://www.dtsc.ca.gov/pollutionprevention/ppac-metal-plating-report.pdf (accessed December 14, 2004).

Carraro, Carlo, and Francois Leveque. 1999. Introduction: The Rationale and Potential of Voluntary Approaches. In *Voluntary Approaches in Environmental Policy,* edited by Carlo Carraro and Francois Leveque. Boston: Kluwer Academic Publishers, 1–15.

Coglianese, Cary, and Laurie K. Allen. 2004. Does Consensus Make Common Sense? *Environment* 46(1): 10–24.

Cooperative Agreement between the City of Statesville and [Metal Finishing Company]. No date. www.strategicgoals.org/tools/localsgp.htm (accessed February 2, 2005).

Duke, L. Donald. 1994. Hazardous Waste Minimization: Is It Taking Root in U.S. Industry? Waste Minimization in Metal Finishing Facilities of the San Francisco Bay Area, California. *Waste Management* 14(1): 49–59.

Duke, L. Donald, Matthew Buffleben, and Lisette A. Bauersachs. 1998. Pollutants in Storm Water from Metal Plating Facilities, Los Angeles, California. *Waste Management* 18: 25–38.

Finishers Management. 2001. Survey on Environmental Compliance 2001. *Finishers Management,* Nov.–Dec. 2001. Formerly available at www.finishers-management.com/novdec2001/sofi2002.htm (accessed December 1, 2003).

Gunningham, Neil, Dorothy Thornton, and Robert Kagan. 2004. Motivating Management: Corporate Compliance with Safety, Health and Environmental Regulation. Working paper 30 (September). Canberra: Australian National University, National Research Centre for OHS Regulation.

Hahn, Robert W., and Gordon L. Hester. 1989. Marketable Permits: Lessons for Theory and Practice. *Ecology Law Quarterly* 16: 361–406.

Heckler, Phil, and Robert LaGrotta. 2000. SGP—A Municipal Perspective. 30 (4) *Clearwaters.* http://www.nywea.org/clearwaters/pre02fall/304050.html (accessed March 2, 2006).

Howard, Jennifer, Jennifer Nash, and John Ehrenfeld. 2000. Standard or Smokescreen?

Implementation of a Voluntary Environmental Code. *California Management Review* 42(2): 63–82.

Jalley, Elizabeth M., Peter Bryan Moores, Brian L. Henninger, and Goud P. Maragani. 2002. Environmental Crimes. *American Criminal Law Review* 39(1): 403–86.

Johnston, Jason Scott. 2005a. Signaling Social Responsibility: Law, Activists, and Market Incentives for Corporate Environmental Performance. Working paper 05-16. Philadelphia: Institute for Law and Economics Research Paper.

———. 2005b. Tradeable Pollution Permits and the Regulatory Game. In *Thirty Years of Market Based Instruments for Environmental Protection: An Assessment*, edited by Jody Freeman and Charles Kolstad. Oxford: Oxford UP.

Kagan, Robert A., Neil Gunningham, and Dorothy Thornton. 2003. Explaining Corporate Environmental Performance: How Does Regulation Matter? *Law and Society Review* 37(1): 51–90.

King, Andrew A., and Michael J. Lenox. 2000. Industry Self-Regulation without Sanctions: The Chemical Industry Responsible Care Program. *Academy of Management Journal* 43(4): 698–716.

Lyon, Thomas P., and John W. Maxwell. 2004. *Corporate Environmentalism and Public Policy*. New York: Cambridge UP.

Maxwell, John W., Thomas P. Lyon, and Steven C. Hackett. 2000. Self-Regulation and Social Welfare: The Political Economy of Corporate Environmentalism. *Journal of Law and Economics* 43(2): 583–618.

Michigan Metal Finishing Strategic Goals Program. 2000. Program Description. www.strategicgoals.org/tools/localsgp.htm (accessed February 2, 2005).

Millimet, Daniel L. 2003. Environmental Abatement Costs and Establishment Size. *Contemporary Economic Policy* 21: 281–96.

MOEA (Minnesota Office of Environmental Assistance). 1999. Metal Plating and Finishing. http://www.moea.state.mn.us/publications/SIC3471.pdf (accessed February 2, 2005).

National Metal Finishing Strategic Goals Program. 2001. Indiana Implementation Plan for 2001–2002. www.in.gov/idem/strategicgoals/implementation.htm (accessed February 2, 2005).

NCMS (National Center for Manufacturing Sciences). 2000. *Benchmarking Metal Finishing*. Report 0076RE00. June. Washington, DC: NCMS.

New York State Strategic Goals Program. 2000. "Metal Finishing" Program Framework. www.strategicgoals.org/tools/PDF/Nyframe.pdf (accessed February 2, 2005).

Nichols, Albert L. 1997. Lead in Gasoline. In *Economic Analyses at EPA: Assessing Regulatory Impact*, edited by Richard D. Morgenstern. Washington, DC: Resources for the Future.

Nixon, Will. 1998. The Thin Green Line. *The Amicus Journal* 20(3): 14–16.

Palmer, Karen, Wallace Oates, and Paul Portney. 1995. Tightening Environmental Standards: The Benefit-Cost or the No-Cost Paradigm? *Journal of Economic Perspectives* 9(4): 119–32.

Porter, Michael E. 1991. America's Green Strategy. *Scientific American* 264(4): 168.

Porter, Michael E., and Claas van der Linde. 1995. Toward a New Conception of the Environment-Competitiveness Relationship. *Journal of Economic Perspectives* 9(4): 97–118.

Schwartz, Richard E., Steven P. Quarles, and Ellen B. Steen. 1999. Encouraging Self-Auditing within the Pork Industry: The Nationwide Clean Water Act Enforcement Agreement for Agriculture's First Industry-Wide Auditing Program. *Environmental Law Reporter* 29: 10395–410.

Segerson, Kathleen, and Thomas J. Miceli. 1998. Voluntary Environmental Agreements: Good or Bad News for Environmental Protection? *Journal of Environmental Economics and Management* 36: 109–30.

Small, Arthur A. III, and Joshua Graff Zivin. 2002. A Modigliani-Miller Theory of Corporate Social Responsibility. Mimeo, Columbia University School of International and Public Affairs and Earth Institute. Available to SSRN subscribers at http://papers.ssrn.com/sol3/papers.cfm?abstract_id=325921.

Speir, Jerry. 2001. EMSs and Tiered Regulation: Getting the Deal Right. In *Regulating from the Inside: Can Environmental Management Systems Achieve Policy Goals?* edited by Cary Coglianese and Jennifer Nash. Washington, DC: Resources for the Future, 198–219.

Stoll, Richard G. 1989. Coping with the RCRA Hazardous Waste System: A Few Practical Points for Fun and Profit. *Environmental Hazards* 1: 6–10.

Strategic Goals Program. 2001a. 2001 Reports: Summary of Progress Report Performance. http://www.strategicgoals.org/reports2/review2a.cfm?state=all (accessed December 14, 2004).

———. 2001b. 2001 Reports: Sludge Generation Normalized Using Sales (Based on Dry Solids). http://www.strategicgoals.org/reports2/t5.cfm?state=all (accessed December 14, 2004).

———. 2001c. 2001 Reports: Toxic Organic Chemical Emissions Normalized by Sales. http://www.strategicgoals.org/reports2/t8.cfm?state=all (accessed December 14, 2004).

———. 2001d. 2001 Reports: Water Discharge Normalized Using Sales. http://www.strategicgoals.org/reports2/t4.cfm?state=all (accessed December 14, 2004).

U.S. Census Bureau. 1999. *Electroplating, Plating, Polishing, Anodizing, and Coloring: 1997 Economic Census.* EC97M-3328C. August. Washington, DC: U.S. Department of Commerce. http://www.census.gov/prod/ec97/97m3328c.pdf (accessed March 6, 2006).

U.S. EPA (Environmental Protection Agency). 1980. Proposed Amendments to Final Rules. Electroplating Point Source Category Effluent Guidelines and Standards Pretreatment Standards for Existing Sources. *Federal Register* 45: 45322, July 3.

———. 1994. Office of Policy, Planning, and Evaluation. Sustainable Industry Project Industry Sectors Report. March 17: 24–27.

———. 1995a. Profile of the Fabricated Metal Products Industry. U.S. EPA Office of Compliance Sector Notebook Project. EPA/310-R-95-007. Washington, DC: U.S. EPA.

———. 1995b. Proposed Rule. Effluent Limitations Guidelines, Pretreatment Standards, and New Source Performance Standards: Metal Products and Machinery. *Federal Register* 60: 28210, May 30.

———. 1998. *Metal Finishing F006 Benchmark Study.* Washington, DC: U.S. EPA.

———. 1999. Common Sense Initiative Metal Finishing Sector Subcommittee Factsheet. http://www.epa.gov/ispd/pdf/pubs_csifactsheet.pdf (accessed December 14, 2004).

———. 2000. Final Rule. 180-Day Accumulation Time Under RCRA for Waste Water Treatment Sludges From the Metal Finishing Industry. *Federal Register* 65: 12378, March 8.

———. 2001a. Proposed Rule. Effluent Limitations Guidelines, Pretreatment Standards, and New Source Performance Standards: Metal Products and Machinery Point Source Category. *Federal Register* 66: 423, January 3.

———. 2001b. *Living the Vision: Accomplishments of the National Metal Finishing Strategic Goals Program.* EPA 240-R-00-007, January.

———. 2002. Notice of Data Availability. Effluent Limitations Guidelines, Pretreatment Standards, and New Source Performance Standards for the Metal Products and Machinery Point Source Category. *Federal Register* 67: 38752, June 5.

———. 2003. Final Rule. Effluent Limitations Guidelines and New Source Performance Standards for the Metal Products and Machinery Point Source Category. *Federal Register* 68: 25686, May 13.

———. 2004. Metal Finishing, http://www.epa.gov/sectors/pdf/metalfinishingbw.pdf (accessed March 17, 2006.

———. 2005. Code of Federal Regulations Title 40 Section 403.8. General Pretreatment Regulations for Existing and New Sources of Pollution, July 1.

Videras, Julio, and Anna Alberini. 2000. The Appeal of Voluntary Environmental Programs: Which Firms Participate and Why? *Contemporary Economic Policy* 18(4): 449–61.

Washington State Department of Ecology. 1996. Pollution Prevention Progress for 23 Electroplating Facilities: An Industry Sector Report, Publication No. 96-426 (August). http://www.ecy.wa.gov/pubs/96426.pdf. (accessed December 14, 2004).

8

Clean Charles 2005 Initiative

Why the "Success"?

Tapas K. Ray and Kathleen Segerson

The Clean Charles 2005 Initiative (CCI), led by the New England office of the U.S. Environmental Protection Agency (EPA) in cooperation with 10 communities that border the Charles River, established the ambitious goal of making the Lower Charles River fully fishable and swimmable by the year 2005. In its initial press release, EPA laid out an eight-point plan toward achieving this goal, which "relies on cooperative efforts among the federal, state and local governments; citizen participation; good science; and, where necessary, strong enforcement" (U.S. EPA 1995, 2). It embodied a series of activities designed to improve water quality in the Charles, including combined sewer overflow (CSO) controls, removal of illegal sanitary connections, stormwater management planning and implementation, public outreach, education, monitoring, enforcement, and technical assistance (U.S. EPA 2000b).

A key component of the initiative is an annual "report card," which provides a grade indicating the progress that has been made in meeting the project's goal, measured in terms of the percentage of time that the water quality standards for bacteria were met. The initial grade given in 1995 was a D. Over the next few years, the annual grades continually improved, reaching a grade of B for 2000. Thus, there was significant improvement in the water quality of the river during this time. Improvement then generally leveled off, as the river continued to receive a B grade in 2001 and 2002, and even dropped to a B– in 2003. Despite EPA's goal of bringing water quality to the level of an A by 2005, the final grade the Charles River received that year was only a B+.[1]

Even though the Clean Charles 2005 Initiative did not achieve its own ambitious goal, it is nevertheless clear that it has achieved significant water quality improvements in the Lower Charles. What is less clear is the explanation for this success. Metzenbaum (2001) argues that much of the success is attributable to the use of an outcome-focused performance goal, where the goal was used to "inspire and inform," "engage media," "enlist allies," "align expectations," and "build collaboration and shared learning." This suggests that the success of the initiative was attributable primarily to its ability to mobilize a series of management-based voluntary efforts to improve the water quality in the river. In fact, Metzenbaum states that "[t]he Clean Charles 2005 initiative demonstrates how the establishment of a clear and resonant goal and timetable, complemented by credible outcome-focused performance measurement, can be a powerful tool for enlisting and engaging the cooperation of [numerous] parties [working together] in the achievement of that goal" (2001, *30*). Is the CCI then an example of successful "leveraging of the private sector," under which government used the management tools of monitoring and information sharing to rally public and private parties to come together cooperatively to meet an environmental challenge? Is it an example of a successful voluntary management-based approach, which demonstrates the desirability of increased reliance on such voluntary approaches and a diminished role for traditional regulation?

This chapter examines the success of the Clean Charles 2005 Initiative, with particular emphasis on what, if any, lessons can be learned from it. We rely both on the recent theoretical literature on the use of voluntary approaches to environmental protection that has emerged within the field of environmental economics and on preliminary empirical evidence relating to the CCI.[2] We conclude that the success of the CCI does not suggest a diminished role for traditional environmental regulatory approaches established under the Clean Water Act (CWA). Rather, we argue that threatened and actual enforcement of existing, traditional regulations played a significant role in the initiative's success. EPA's use of management-based strategies—in particular, its establishment of the goal of a swimmable and fishable Charles River by 2005—undoubtedly motivated numerous parties, including EPA staff.[3] However, the existence of a traditional regulatory structure that could be used to threaten or actually impose financial penalty or injunction appears to have provided a significant part of the "motivation" for private parties to contribute toward meeting the goal through reduced discharges to the river. Had this regulatory structure not been in place, it is likely that the initiative would have been less successful. In fact, the difficulty that EPA now faces in achieving further water quality improvements (i.e., in moving to an A grade) seems to stem at least in part from the fact that the remaining pollution sources (primarily non-point sources of pollution) are not easily regulated under traditional approaches and hence more difficult to control (U.S. EPA 2001).

The chapter is organized as follows. We begin with brief background descriptions of water quality in the Charles River and the activities that EPA undertook as part of the initiative to clean it up. We then turn to the existing literature to identify the factors that are likely to make an effort of this type successful, and use the insights from this literature to develop a conceptual framework for analyzing the impact of such a program and identifying those components that appear to be most responsible for its success. Finally, we present some preliminary empirical results regarding factors that have affected water quality in the Lower Charles since the initiative began in 1995. We use both the theoretical literature and the empirical results to argue that regulatory enforcement played a significant role in the CCI's success. We do not, however, suggest that the CCI had little effect. Rather, it is our contention that the initiative paved the way for a successful enforcement effort.

Water Quality in the Charles

The Charles River watershed is one of the most populated watersheds in New England. It extends inland from Boston Harbor, encompassing 311 square miles (U.S. EPA 1999c). The watershed consists of three portions: Upper, Middle, and Lower Charles (DEP 2000). The primary focus of the CCI is on the Lower Charles basin, although some aspects of it relate to contamination sources in both the Upper and the Middle Charles. Our analysis is focused primarily on cleanup efforts in the Lower Charles region.

For purposes of the CCI, water quality in the river is based on the number of colony-forming units of fecal coliform bacteria present in 100 milliliter water samples.[4] The water samples used for testing are collected monthly at sampling stations along the Charles under the guidance of the Charles River Watershed Association (CRWA).[5] To be fit for swimming, the bacterial count must not exceed 200 colony-forming units per 100 milliliters of water, while the maximum level for boating is 1,000.

The major sources of bacterial pollution in the Lower Charles River include both point and non-point sources. The primary point sources of bacteria are stormwater drains, illegal sewer connections to the stormwater drains, CSOs, and wastewater treatment plants. The wastewater treatment plants are mainly a problem in the Upper Charles, where these aged plants sometimes fail to remove bacterial content from the discharged water (DEP 2000). In the Lower Charles Basin, on the other hand, wastewater is collected by the municipalities through various sewer networks. This wastewater is then transported to offsite treatment plants near the Boston Harbor through the CSOs.

In addition to carrying wastewater to treatment plants, CSOs also function as storm sewers. Runoff from streets is carried in the same pipes as sanitary sewage. During normal weather, the wastewater treatment plants can process this combined waste stream. However, during heavy rainstorms, the capacity of the pipes carrying the wastewater to the treatment facility is exceeded, causing potential flooding. As a result, rainwater runoff mixed with the sewage is often discharged directly into the Charles River and Boston Harbor through the CSOs. In 1995, there were 19 permitted CSOs in the Lower Charles, only one

TABLE 8-1. River Grade and Percentage of Days River Met Standards

Year	Wet Swimming	Wet Boating	Dry Swimming	Dry Boating	Overall Swimming	Overall Boating	Grade
1995	—	—	—	—	19	39	D
1996	15	45	40	94	21	57	C-
1997	22	61	56	87	34	70	C
1998	31	74	85	98	51	83	B-
1999	62	85	71	100	65	90	B-
2000	46	91	82	94	59	92	B
2001	40	74	80	87	54	82	B
2002	21	86	71	100	39	91	B

of which treated its waste before discharge (DEP 2000). Apart from the CSOs there are numerous stormwater drains that empty into the Charles River, some of which are illegally hooked up to sewer lines or combined sewer lines (*Boston Globe* 1996).

The primary non-point source of pollution to the river is stormwater runoff, which carries bacteria from various sources including animal waste. Studies done by EPA have shown that rainfall of 1 inch has the potential to wash more than 100,000 pounds of solids into the Lower Charles (U.S. EPA 1995). According to EPA's press release in October, 1995, "as much as eight thousand pounds of nitrogen and phosphorus, more than six hundred pounds of copper and zinc, and large quantities of harmful bacteria enter the river through stormwater runoff" (U.S. EPA 1995, *1*).

The bacterial count depends not only on pollution sources but also on rainfall, water flow, and temperature. An increase in count occurs during wet weather. To account for this, EPA assigns two separate grades—one for wet weather and one for dry weather—as well as an overall grade that combines the two. The grades that have been assigned since the program's inception are provided in Table 8-1. This table also shows the percentage of days the river was fit for primary and secondary contact in any given year.

Figure 8-1, which reports the actual bacterial count at one monitoring station (729S, located at Eliot Bridge, Cambridge), indicates the monthly variability in bacterial count. The two highest spikes (in October 1996 and August 1998) correspond to heavy rainfalls, although a comparison with rainfall data shows that such rainfall events were not always associated with bacterial pulses of this type. It is clear from this figure that the bacterial count varies considerably over time. Other monitoring stations exhibit similar patterns.[6]

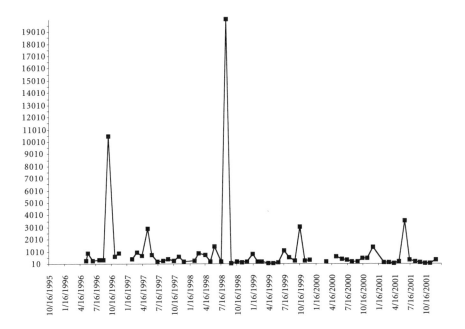

FIGURE 8-1. Monthly Bacterial Count for the Monitoring Station 729S

Activities under the CCI

Several different organizations, both EPA and nongovernmental, have worked together to clean up the Charles River since the initiative began.

EPA Activities

As noted above, the Clean Charles 2005 Initiative started with an EPA proclamation to the press on October 22, 1995, stating that the long-contaminated Lower Charles River would be clean enough for fishing and swimming by 2005. At the time of the initial announcement, EPA had already begun enforcement against illegal discharges of sewage to storm drains along the river. In fact, the idea for the CCI stemmed from two enforcement cases at EPA, one against the Massachusetts Water Resource Authority (MWRA) relating to the control of CSO discharges, and the other against the town of Brookline relating to sewer connections to stormwater drains.[7]

The significant components of the initiative included (1) a performance goal for the river (fishable and swimmable by 2005), (2) a mechanism for regularly monitoring and reporting progress toward the goal (the "report card"), and (3) a comprehensive plan for achieving the goal. This plan was initially based on an eight-pronged approach (U.S. EPA 1995):

- conventional EPA enforcement to ensure removal of illegal sewage discharge to storm drains,
- implementation of comprehensive stormwater management programs,
- formation of "Task Force 2005"—the Clean Charles Steering Committee,
- EPA establishment of the Charles River Enforcement Team,
- control of combined sewer overflows,
- support for the Charles River Watershed Association's IM3 (Integrated Monitoring, Modeling, and Management) study,
- development of funding sources, and
- Metropolitan District Commission participation.[8]

With this plan in hand, EPA began implementation. It undertook an enforcement initiative against illegal discharges of sewage to storm drains along the Charles. By order of EPA, Cambridge, Brookline, Watertown, Newton, Waltham, Needham, Wellesley, Weston, and Dedham were supposed to locate and remove illegal discharges by 1997 (Brelis 1995; Howe 1996).[9] In 1996, EPA investigators extensively surveyed the riverbanks and discovered more than 300 illegal sewer discharges. Eliminating these discharges stopped 20 million gallons of sewage each year from entering the river (U.S. EPA 1997). By 1998, "all 10 communities in the Lower Charles basin had evaluated their storm drains for illegal tie-ins from sewer pipes and removed connections, reducing illicit discharges by one million gallons a day" (U.S. EPA 2000a). EPA ordered Milford to shut down all illegal sewer connections to storm drains by November 1997 or face possible fines or lawsuits (Howe 1996).

EPA also sought to have comprehensive stormwater management programs in operation in all communities on the Lower Charles by December 1996. In 1997, under the threat of CWA lawsuits, EPA required public works agencies in 10 communities bordering the river from Dedham to Boston to file plans by the following July detailing how they would improve street sweeping, catch-basin cleaning, and sewer maintenance, and how they would keep pollution from storm drains (U.S. EPA 1997).

Enforcement actions were also taken in the Upper Charles Region. The six wastewater treatment plants were now continuously monitored, and they faced stricter permit requirements than before. Enforcement actions were taken against the Charles River Pollution Control District (CRPCD) in Medway for its failure to implement adequate pretreatment of industrial wastes as required by federal discharge permits (U.S. EPA 1999a; 1999b).

Punitive actions were also being taken against facilities that caused a potential threat to the river quality in general. In 1995, EPA fined Conrail for illegal discharges into the Charles (Wald 1995). In 1996, one of the largest-ever federal environmental penalties against an educational institution was levied against Boston University for two oil spills into the Charles River (Gelastopoulos 1997). In 1998, EPA issued a formal CWA demand letter to the Cambridge

Electric Company, which was identified as the probable source of recurring oil sheen in the Charles. In 1999, EPA fined Watertown for a violation of the CWA at its public garage.

The next year, EPA proposed a $396,299 penalty against the Natick Public Works Department for alleged illegal pollution at a town garage. Northeastern University was fined $20,000 for violations detected by environmental regulators at a dormitory construction site. Out of this amount, $10,000 was paid to the CRWA to study pollutants in Charles River stormwater runoff. In 2000, EPA fined Boston Sand and Gravel for illegally discharging wastewater over the span of several years into the Charles River basin and two other sites. The $1.3 million fine represents one of the largest fines ever imposed for violating the CWA. In 2001, MIT agreed to pay a $150,000 federal fine for hazardous waste violations. This settlement stemmed from 18 violations that EPA inspectors found in May 1998 at the Cambridge campus.[10]

Often enforcement actions could be avoided by prior warnings that worked as background threats. These threatened enforcements led to inducement of "voluntary" reductions in discharges to the river (U.S. EPA 1998). For example, in March 1998, EPA warned 200 public and private facilities situated in the vicinity of the river that it would undertake an enforcement sweep. This, coupled with actual unannounced inspection of the facilities on May 1, 1998, found very few detected violations of state and federal laws. John DeVillars, chief of EPA Region 1 at that time, commented that "It appears our warning letters worked very well in persuading companies along the river to make sure their operations were clean and up to the code" (U.S. EPA 1999d, *2*).

In addition to stepping up enforcement actions, EPA also sought to involve the various stakeholders through forming task forces and providing funds and technical assistance to different environmental and citizen groups working in the Lower Charles (U.S. EPA 1998).[11] A portion of the money EPA collects by penalizing violators is earmarked for funding various projects carried out by different environmental groups. There are also instances where EPA has provided subsidies to the neighboring communities in the Lower Charles in order to assist them in solving illegal sewer connection problems. For example, EPA, along with the Massachusetts Department of Environmental Protection (DEP), announced that $25 million was available in interest-free loans to municipalities in combined federal and state dollars to fund various abatement activities on the Charles (U.S. EPA 1998). EPA also provides technical assistance to municipalities and communities in their ongoing efforts to plan stormwater improvements. In May 1998, EPA provided assistance to more than 1,000 small businesses around the Charles to maintain stormwater management plans in their garages (U.S. EPA 1998). Along with the DEP, EPA has developed funding strategies to support local communities' efforts toward stormwater control.

Considerable emphasis has been put on disclosing information about the initiative's progress and on the activities responsible for it. Providing newspaper reports, stenciling drains,[12] announcing the river grade during the popular boating race known as the "Run of the Charles," setting up a hotline for public complaints, and educating school children and citizen groups all may have led to greater involvement from various stakeholder groups. In 1998, the CRWA (with help from EPA) initiated a water quality flagging program, which produces real time data about water quality conditions for a stretch of the river in its lower basin. Working with the Natural Resource Conservation Service, EPA has also created a stormwater education handbook that towns and cities can use to educate their citizens about the potential damage of runoff and how citizens can reduce this damage. EPA is continuing to hold workshops with municipalities on how to implement these ideas.

In 1998, EPA's Office of Environmental Measurement and Evaluation established a water quality monitoring program (the Core Monitoring Program) in the Charles River. The purpose of this program is to track water quality improvements in the Charles River Basin and to identify where further pollution reduction or remediation actions are necessary in order to meet the swimmable and fishable goals.

MWRA Activities

Apart from EPA, there have been other contributors to the Clean Charles 2005 Initiative from both the public and private sectors. For example, the MWRA collects, treats, and discharges municipal wastewater from the Lower Charles to the Boston Harbor. The authority was created in 1985 with the sole purpose of providing clean water and sewage treatment in the Lower Charles area.[13] The MWRA Combined Sewer Overflow Control Plan found that stormwater loads are the predominant source of pollutants in the Lower basin. It developed the CSO Action Plan, the primary objective of which was to reduce the number of untreated CSO facilities. A study done on behalf of the MWRA has shown that the CSO Action Plan reduced bacterial counts during the period up to 1995 (MWRA undated). However, there does not appear to be any published work showing the effect of the CSO Action Plan on bacterial count in the post-1995 period. Out of the 19 permitted CSOs that were discharging in this region, only 11 now discharge into the Lower Charles (Gray et al. 2003). Nearly 1.5 billion gallons a year of treated discharge have been eliminated . This is largely due to the improved pumping capacity at the Deer Island Wastewater Treatment Plant. The Cottage Farm Treatment Plant has also been upgraded and work has begun to upgrade the Prison Point Facility.

Private Activities

Foremost among the nongovernmental organizations is the Charles River

Watershed Association (CRWA). Founded in 1965 in response to public concern about the declining condition of the Charles River, the CRWA takes pride in being one of the country's first watershed organizations. Since its earliest days of advocacy, the CRWA has figured prominently in major cleanup and watershed protection efforts, working with government officials and citizen groups from 35 Massachusetts watershed towns from Hopkinton to Boston. Since 1995 the CRWA has monitored river quality throughout its entire length and thus has made it possible to identify sudden abnormalities caused by illegal polluting activities. This has greatly helped in improving the river quality. The CRWA acts as an advisory body to the towns about watershed management and is involved in various disclosure programs such as flagging river quality and publishing newsletters. Among its programs is Smart Storm, a stormwater management system that helps in the retention of rainwater for various domestic uses.

Another nongovernmental organization that has been actively involved in CCI is the Clean Charles Coalition, a voluntary association of industries, academic and research institutions, concerned citizens (including people owning land adjacent to the Charles), public interest groups, and other organizations that have joined in support of a fishable and swimmable Lower Charles River by 2005. In 1999, the group comprised 11 landowners; by 2002, the number of group members had increased to 15 (Gray et al. 2003). These members finance educational and public outreach programs and undertake voluntary actions to improve water quality. The main agenda is to work together on best management practices (BMPs) to control stormwater pollution. Coalition members are expected to make an active contribution to the coalition's goals, including participating in scheduled meetings and contributing at least 100 hours of volunteer time each year to Coalition projects.

Relative Contributions of Enforcement Actions

The above summary suggests that the Clean Charles 2005 Initiative has included both enforcement actions (actual or threatened) and an appeal to voluntary management-based approaches that emphasize monitoring and information sharing. What were the relative contributions of the enforcement-related actions and those actions that might have been undertaken on a purely voluntary basis? Although we are not able to measure directly the impact of truly voluntary efforts, we argue below that a significant contribution to date in improving water quality in the Lower Charles River has come from enforcement actions, most notably the elimination of illegal sewer hookups and the control of CSOs. In the following sections, we present theoretical and preliminary empirical support for this conjecture.

What Does the Literature Suggest?

A key component of the Clean Charles 2005 Initiative was the establishment of a performance goal. Performance goals can serve to motivate certain types of behavior. It is important to distinguish between motivation provided to the organization's employees (in this case, EPA staff) and motivation provided to others outside the organization. The literature on the use of performance goals has focused primarily on the former group. Previous studies of the use of performance measures by a private or public organization to motivate employees have found that they can increase incentives for employees to work harder to achieve those goals (e.g., Holmstrom 1979).

However, since true performance is often difficult to measure, particularly in the context of public agencies, use of performance measures can also provide incentives for employees to "game the system" in an effort to achieve their own private goals, when those goals are not exactly aligned with the goals of the organization or agency (Baker 1992; Courty and Marschke 1997; Heckman et al. 1997). In addition, performance goals are likely to work more effectively when goals can be clearly specified and quantified (e.g., Heckman et al. 1997), when workers participate in the development of the goals (Christensen 1982), when positive feedback regarding progress toward the goal is provided (Foran and Decoster 1974), and when workers are motivated to work cooperatively toward the goal rather than competitively.

It appears that the CCI met these criteria. The goal was clearly specified, EPA staff participated in the development of the goal, and the annual report card provided a mechanism for positive feedback on progress. Furthermore, the goal was a collective goal rather than a series of (competitive) individual goals. In addition, Metzenbaum (2001, *10–11*) argues that the performance goal under the CCI was effective because it was "simple, resonant, and understandable; outcome-focused; time and location specific; ambitious but feasible, and public."

It seems clear that the CCI's performance goal motivated staff within EPA to strive to achieve the goal.[14] The question then is whether, or to what extent, the goal also motivated others outside the organization to undertake voluntary actions to contribute to the goal. There is a growing literature on the effectiveness of environmental programs that rely on voluntary actions to achieve their goals.[15] This literature suggests that, in order for a voluntary approach to be successful, there must be adequate incentives for participation.

What motivates participation in voluntary environmental programs? Participation incentives can stem from at least three different sources: environmental stewardship, market-based incentives, and government-created incentives (see, e.g., Alberini and Segerson 2002). Consider first the role of environmental stewardship motivated by personal satisfaction gained from undertaking activities that protect the environment. If this motivation exists and is sufficiently strong, then increased education and awareness can lead to changes in behavior that reduce environmental degradation. This motivation is more likely to exist in cases where pollution stems from individual behavior (such as lawn fertilizing) and the cost of a behavioral change is relatively small. However, even in these cases incentives may be weak if individuals feel that their contribution to the aggregate pollution loading is small and hence that changes in their behavior are unlikely to have a noticeable effect on environmental quality.

Organizations such as firms, educational institutions, or public organizations are less likely to be motivated directly by environmental stewardship, since providing environmental quality is not generally their main objective. However, these organizations can be motivated indirectly by the environmental stewardship goals of those they seek to serve or the communities in which they are located. For example, in the private sector it may be possible for consumers with "green preferences" to induce the use of environmentally friendly production processes or the production of environmentally friendly products through the marketplace. If there is sufficient demand for green products, firms can stand out among their competitors if they supply this niche. With sufficient demand, exploiting the "green" market can increase market share and firm profits, even when production costs for green products are higher. Alternatively, firms that produce intermediate goods might be motivated to undertake voluntary environmental protection in an effort to improve access to or terms received in input markets. In either case, the incentive for voluntary abatement can be provided by the market. Market-based incentives, however, apply primarily to the private sector. Although public sector or not-for-profit polluters may be responsive to input markets, they do not generally sell goods or services directly to consumers in output markets.

In the absence of strong market incentives, regulatory agencies may be able to induce voluntary pollution abatement by creating either positive ("carrot") or negative ("stick") incentives for polluters to participate in voluntary programs. Positive incentives usually take the form of financial subsidies or cost-sharing schemes; negative incentives take the form of explicit or implicit threats to impose a costly outcome on polluters if sufficient abatement is not undertaken "voluntarily." This outcome may be the imposition of a direct financial penalty or the withholding of some otherwise favorable treatment. It might also entail the imposition of more costly regulation, such as more stringent or less flexible permit requirements.

Positive incentives work by reducing the polluters' costs and hence increasing the likelihood that they will be willing to undertake pollution abatement voluntarily. Although empirical evidence suggests that subsidies can be effective in inducing voluntary abatement,[16] subsidies can be socially costly because of the tax-related distortions that stem from the need to raise the revenue to finance the subsidy.

In contrast, negative incentives do not require a financial outlay by the regulatory agency. They work by making failure to act "voluntarily" more costly than taking voluntary action. However, if the negative incentives take the form of threats, the threat must be credible—that is, the regulator must be willing to follow through on it if necessary. The threat of enforcing an existing regulation if pollution is not reduced voluntarily is generally a more credible threat than the threat of enacting new regulations, since it is more likely to be imposed (Segerson and Miceli 1998).

Based on their review of both the theoretical and empirical literature on voluntary approaches to pollution control, Alberini and Segerson (2002) identify two key features that are likely to increase the effectiveness of voluntary approaches. The first is the existence of a strong regulatory threat. Such a threat can provide strong incentives for participation, particularly when it takes the form of threatening to enforce existing regulations. The second feature is the availability of reliable monitoring. Since regulatory threats require imposing penalties if and when the voluntary approach fails to achieve environmental targets (either individually or collectively), a mechanism for reliably determining whether those targets have been met must be available. When based on a strong and credible regulatory threat and a reliable means of monitoring success, a voluntary approach can provide a mechanism for effectively meeting environmental quality goals (Alberini and Segerson 2002). However, absent such a threat, a voluntary approach is likely to be successful only if environmental stewardship or the existence of a "green demand" is sufficiently strong, or financial inducements to offset some share of abatement costs are provided.

Although the literature on voluntary approaches focuses primarily on the incentives for participation and the likelihood of success, we can use the insights from this literature to examine the question of interest here, namely, the various channels through which a performance goal might motivate others outside of the organization to contribute to the goal. Figure 8-2 presents a framework that summarizes these various channels, which we describe below. It draws from the literature relating to performance goals as well as the economic literature on voluntary approaches to environmental protection.

What does the literature suggest about the possible impacts of a performance standard? As indicated in Figure 8-2, first and foremost, establishing and publicizing a performance goal increases awareness among a variety of stakeholder groups, including regulators, residents, environmentalists, community groups, regional or local governmental and quasi-governmental officials, and firms and businesses operating in the area. With increased awareness comes the potential for increased accountability, particularly if the group responsible for meeting the goal is clearly identified. Increased awareness and accountability can contribute to an increased resolve to improve environmental quality in an effort to meet the performance goal.

At least four potential effects of increased awareness of and accountability for the environmental quality of a given area exist. One key effect is an increased

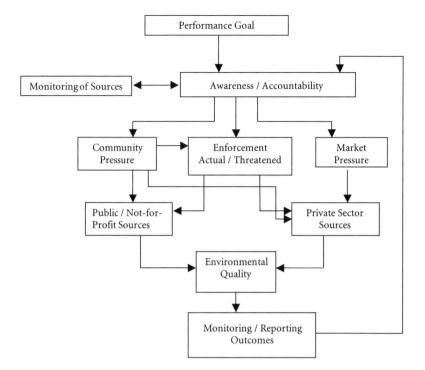

FIGURE 8-2. Channels through which Setting Performance Goals Affects Environmental Quality

effort to identify and monitor pollution sources. The information generated by this monitoring further increases awareness, in particular awareness of individual contributions. Awareness is thus likely to grow over time, and its increase can in turn lead to three other effects: increased community pressure, increased enforcement, and increased market pressures.

As residents who live in the vicinity of the polluted waterbody become more aware of the environmental degradation, they are likely to put pressure on all polluters to take steps to reduce the amount of pollution in the waterbody. These polluters could include not only firms (i.e., the private sector)[17] but also other public or not-for-profit sources such as local municipalities, educational institutions, and households (from fertilizers, sewers, and pets). Community pressure can work through reputations, elections, or peer pressure.

Although community pressure can affect both the private sector and other not-for-profit sources, market pressures are effective only on sources that rely on markets. To the extent that a firm's consumers or input suppliers are concerned about its environmental performance, increased awareness of that performance can increase the incentive for firms to improve performance to avoid losing or alienating consumers or suppliers. If the price at which it can sell its product, the extent of its market, or the cost of its inputs depend on its environmental performance, then firms can be expected to respond to these market pressures.

Finally, increased accountability coupled with the improved information about sources can lead to increased enforcement actions. Both private sector and public or not-for-profit polluters can be expected to respond to enforcement actions if the cost of complying with the relevant regulation is less than the cost of not complying (for example, the penalty for noncompliance). Note that the incentive to comply exists if an enforcement action is not actually undertaken but instead threatened (for example, the firm is told that it will be inspected and an enforcement action will be undertaken if it is found to be noncompliant), provided the firm views the threat as credible. In addition, even if no regulation is in place, the threat of the imposition of a costly regulation (or tax) can also provide an incentive for a firm to reduce pollution "voluntarily." Again, however, in order for such threats to be effective, they must be viewed as credible.

If community, market, and enforcement pressures lead to reductions in private sector and other sources of pollution and hence to improved environmental quality, the monitoring or reporting of this improvement can further increase awareness. Thus feedback on the extent to which the performance standard has or has not been met can play a key role in determining its ultimate effectiveness. The link from monitoring outcomes back to awareness in Figure 8-2 highlights the importance of this feedback.

The framework outlined in Figure 8-2 indicates a number of channels through which a performance standard can lead to improved environmental quality. The description of the activities undertaken by EPA under the CCI suggests that the initiative sought to exploit several of these channels. The literature on voluntary approaches discussed above suggests that a primary channel leading to the success of a "voluntary" initiative is likely to be through regulatory threats. Consistent with this, we argue below that, in the case of the Clean Charles Initiative, a primary channel that has been operative to date is the effect of increased awareness and accountability on enforcement of regulations relating to noncorporate sources of pollution (most notably, illegal sewer hookups and CSOs). Thus, we argue that the channel outlined in bold in Figure 8-2 played a key role in the CCI's success.

What Does the Empirical Evidence Suggest?

In an effort to assess the role of enforcement actions and CSOs in reducing pollution in the Lower Charles, we conducted a simple regression analysis. Such an analysis can provide information about the extent to which water quality improvements can be explained by actions related to enforcement and CSOs.

TABLE 8-2. Change in Bacterial Count over Time
Dependent Variable: Log (Bac_Count)

Variables	GLS estimates	Standard error
Constant	5.809	0.192
Year 97	− 0.392	0.264
Year 98	− 0.446	0.266
Year 99	−0.773	0.249
Year 00	−0.438	0.267
Year 01	− 0.809	0.254
Year 02	− 0.173	0.264

$R^2 = 0.03$; number of observations 594; degrees of freedom 14.

Because we do not have data on voluntary pollution reductions, we are not able to assess the relative contribution of the components of the CCI that are targeted toward these reductions. Likewise, we are not able to observe the role that *threatened* enforcement played in those cases where polluters responded to the threat, thereby eliminating the need to undertake an actual enforcement action. Nonetheless, our analysis shows that a significant part of the improvement that has been observed over time can be explained by enforcement actions and the closure of CSOs.

We assume that water quality in the river is a function of stream flow, rainfall, the number of CSOs, and the number of enforcements actions relating to bacterial sources undertaken by EPA. The dependent variable in the analysis is the logarithm of the number of colony-forming units of fecal coliform bacteria in 100 milliliters of water. We use monthly time series data from May 1996 to December 2002, although some months were dropped due to an unavailability of monitoring data for those months. The monthly data are based on samples collected by volunteers under the guidance of CRWA. They are available on the CRWA Web site.[18] Our primary focus is on the stretch of the Charles River that flows from the Watertown Dam to the New Charles River Dam. CRWA has nine monitoring stations in this stretch.

With the above dependent variable we run three regressions. The first regression identifies the time trend of the bacterial count starting from 1996. Log transformed bacterial counts are being regressed on year dummies. A positive coefficient implies an incremental increase in bacterial count, while a negative coefficient implies a decline and thereby an improvement in water quality. The results as given in Table 8-2 indicate overall improvement in water quality over these years in comparison with that of the base year and thus validate EPA's improving river grade. However, in both 2000 and 2002 the bacterial count increased relative to the previous year.

Our interest is not simply in showing the reduction in bacterial count but rather in explaining what has contributed to this reduction. In the second and

TABLE 8-3. Descriptive Statistics for Explanatory Variables

Variable	Count	Mean	Maximum	Minimum	Standard Deviation
RAINFALL	594	0.499	2.98	0	0.631
STREAMFLOW	594	5.361	6.899	3.339	0.941
BACT_1	594	5.298	10.352	-0.531	1.624
CSORAIN	594	0.781	14.9	0	1.769
ENFORCEMENT					
ACTIONS	594	0.32	3	0	0.68

third regression models, we include variables that capture the actions that we hypothesize have contributed significantly to improved water quality. In the second regression, we consider data from 1996 to 2002; in the third we consider only the period 1996 to 1999. We consider the two time periods separately because of the higher frequencies of enforcement actions during or before 1999.

We model the dependent variable as a function of the following observed explanatory variables: rainfall, streamflow, the bacterial count in the previous station, the number of active combined sewer overflows, the presence of the Deer Island treatment facility, and the number of punitive actions taken before the monitoring data. In addition we also include (alternative specific) station dummy variables. These capture the effect of unobserved station-specific attributes, including pollution inputs at a site that do not vary considerably over our timeframe (due, for example, to population density or riverbank characteristics). We define each of the explanatory variables below. Table 8-3 provides the descriptive statistics for these variables, except the one for Deer Island, which is simply a dummy variable.

1. RAINFALL: The amount of rainfall at Logan Airport for the three days before the samples are collected and on the date of sampling. Data on the amount of rainfall are available on the CRWA Web site. We constructed a weighted average of these rainfall data. Rainfall is expected to have two effects. The first is a direct effect through runoff of surface pollutants. The second is an effect through CSOs (see below). Both effects are expected to be positive.

2. BACT_1: Logarithmic transformation of bacterial count at the last station divided by the distance between the two stations. At any given monitoring site, the water quality is likely to be affected not only by sources at that site but also by the quality of the upstream water that flows into that site. Thus the coefficient on this term is expected to be positive. The adjustment for distance between sites is intended to capture the dissipation that occurs as the water flows downstream.

3. STREAMFLOW: Logarithmic transformation of the monthly stream flow

data measured in cubic feet, as given in the U.S. Geological Survey Web site. This flow is measured at the surface level of the river. Though the effect of stream flow on bacterial count is not clear, we expect it to be negative, since increased flow implies higher volume and thereby less concentration.

4. ENFORCEMENT ACTIONS: We summed the number of enforcement actions that were directly related to bacterial pollution at a given station. A list of the enforcement actions used to construct this variable is provided in Appendix 8-1. We anticipate that an increase in enforcement actions will reduce the bacterial count. Information on these enforcement actions, along with various other activities in the Lower Charles, was obtained through an extensive search of the leading newspapers, bulletins, and magazines of the region. This search was performed through Lexis-Nexis from the years 1995 to 2003.

5. CSORAIN: We measured the number of CSOs upstream, but downstream to the previous monitoring points, in any given period. Because we expect the impact of CSOs to be greater when rainfall is heavy, we included CSOs using an interaction term with rainfall. We would expect a positive coefficient on this term, which implies that the effect of CSOs on the bacterial count is higher when rainfall is higher.

6. DEER_ISLAND: We used a dummy variable for the Deer Island wastewater treatment plant. This variable was set to 1 for all years after 1998, which is the year in which the wastewater treatment plant began its operation at full capacity. The coefficient on this term is expected to be negative, since greater capacity should reduce direct discharges into the river.

The basic regression equation is:

$$Log[BAC\_COUNT]_{jt} = \alpha + \gamma_j D_j + \beta_1 RAINFALL_t + \beta_2 CSORAIN_{jt} + \beta_3 BACT\_1_{jt}$$
$$+ \beta_4 \sum_{t=0}^{t=t-1} ENFORCEMENT\_ACTIONS_{jt} + \beta_5 STREAMFLOW_{jt} \quad (8.1)$$
$$+ \beta_6 DEER\_ISLAND_t + \varepsilon_{jt}$$

where j denotes the jth monitoring station, t denotes the time period, D_j denotes dummy variables for the stations, and ε_{jt} are unobserved random terms.

Equation (8.1) explains the bacterial count in period (t) as a function of the total number of enforcement actions taken through period (t–1) and other explanatory variables. The objective is to see the effect that these actions had on river pollution. However, it is possible that the enforcement actions in any given period are the result of a high bacterial count in that period, introducing the possibility of endogeneity in the model. To account for possible endogeneity, we also considered a simultaneous equation model, which comprised (8.1) plus the following equation:

$$ENFORCEMENT\_ACTIONS_{jt} \quad = \theta_{jt} + \lambda_1 Log[BAC\_COUNT]_{jt} + \lambda_2 \sum_{t=0}^{\cdot} LAWS_t + \upsilon_t \quad (8.2)$$

Here LAWS represents the number of laws passed before the sampling date that have a direct or indirect effect on the river quality (such as the CWA). However, when we estimated the simultaneous equation model, λ_1 was found to be insignificant and thus we rejected this model in favor of the simple regression in (8.1). The regression results for the simultaneous equation model are given in Appendix 8-2.

Results

Table 8-4 presents the regression results for equation (8.1) for the two models corresponding to the two different time periods (1996–2002 and 1996–1999). The results show that enforcement actions are highly significant (at the 1 percent level) in explaining the reduction in bacterial counts in the early period, during which most of the enforcement activity, as well as most of the improvement in water quality, took place (Model 3). The enforcement variable is significant at the 5 percent level in explaining variability over the full time period (Model 2). The results suggest that enforcement actions were a key contributor to the improvement in water quality. Moreover, the higher magnitude of the coefficient during the period of the CCI's success (1996–1999) suggests that this success was largely due to the enforcement actions taken during that period. In addition, it is likely that our results underestimate the impact of enforcement, since we are not able to capture the effect of *threatened* enforcement by EPA when the threat was successful (and hence an actual enforcement action was never taken). The stronger effect of the enforcement variable in the earlier period is especially interesting, given the qualitative nature of this variable. (We have been able to measure only the number of actions and have not accounted for differences in the nature or severity of these actions.) The strength of the effect suggests that, independent of their severity, the number of enforcement actions may have been viewed as an indicator of the perceived strength of the enforcement threat, which in turn motivated polluters to identify and reduce pollution sources.

Overall, the qualitative results from the two models over the two different time periods are quite consistent. First, both show that, ceteris paribus, the monitoring sites in Boston and Cambridge (stations 4–9) generally have lower bacterial counts than the upstream stations (Watertown, Newton, and Brighton). In addition, both show that the bacterial count in the previous station has a significant positive impact on the next station's count, implying that a significant part of the bacteria at any given station is attributable to upstream sources.

The direct effect of rainfall is also positive and significant, as expected, sug-

TABLE 8-4. Determinants of Bacterial Content in the River
Dependent Variable: Log (Bac_Count)

Variable	Model 2: 1996–2002 $R^2 = 0.427, N = 590$		Model 3: 1996–1999 $R^2 = 0.477, N = 336$	
	Coeff.	t-ratio	Coeff.	t-ratio
INTERCEPT	3.54828*** (0.4326)	8.20	3.68*** (0.549)	6.71
STATION 2	0.0760 (0.220)	0.34	0.215 (0.303)	0.71
STATION 3	0.1019 (0.227)	0.45	0.645** (0.321)	2.01
STATION 4	−0.689*** (0.238)	−2.9	−0.158 (0.322)	−0.49
STATION 5	−0.649*** (0.220)	−2.95	−0.329 (0.307)	−1.07
STATION 6	0.40638* (0.2172)	1.87	0.816*** (0.305)	2.67
STATION 7	−1.703*** (0.267)	−6.37	−1.26*** (0.367)	−3.42
STATION 8	−1.701*** (0.236)	−7.19	−1.70*** (0.326)	−5.22
STATION 9	−1.015*** (0.217)	−4.68	−1.24*** (0.312)	−3.98
RAINFALL	0.44407*** (0.119)	3.70	0.919*** (0.226)	4.07
ENFORCEMENT ACTIONS	−0.0571** (0.027)	−2.07	−0.07*** (0.031)	−2.59
CSO * RAINFALL	0.1604*** (0.044)	3.61	0.123 (0.075)	1.64
DEER ISLAND	0.0148 (0.167)	0.09	0.152 (0.276)	0.55
STREAMFLOW	0.0334 (0.056)	0.60	0.0301 (0.074)	0.41
BACT_1	0.4226*** (0.043)	9.78	0.0432*** (0.059)	5.96

** indicates significance at the 5 percent level; *** indicates significance at the 1 percent level.

gesting that heavy rainfalls that wash pollutants into the river contribute significantly to degraded water quality. Likewise, the results suggest that CSOs contribute significantly to bacterial pollution when rainfall is heavy. Thus the reduction in the number of CSOs that has occurred appears to have been an important factor in the improvements that have been observed, although the effect comes primarily during heavy rainfalls (as expected). Note that the effect of the increased capacity at Deer Island is negative (as expected) but not significant. It is possible that the beneficial impact of this increased capacity is captured through the reduction in the number of CSOs.

In addition to the effect of the individual explanatory variables, it is interesting to look at the overall explanatory power of these variables. The R^2 for the third regression implies that 42 percent of the variability in the reduction in bacterial counts over the full time period is explained by the variables in the model. This suggests that a significant amount of the improvement in water quality that has occurred over time can be explained by factors that are not related to purely voluntary efforts.

Lessons Learned

Based on the description of the activities undertaken under the CCI, the theoretical literature on voluntary approaches to pollution control, and the preliminary statistical evidence presented above, we contend that enforcement actions, either threatened or actual, played a significant role in generating the water quality improvements that were observed under the CCI. It appears that major improvements from both the private sector and public or not-for-profit organizations were in response to actual or threatened enforcement actions, along with the continued efforts on CSOs begun under the CSO Action Plan, rather than a result of EPA's "leveraging" through voluntary management-based strategies.

What role then did the CCI play in achieving the water quality improvements? Could or would these enforcement efforts have been successfully undertaken without the CCI? We believe that the management-based components of CCI, in fact, played a crucial role in the improvement in water quality in the Lower Charles. It appears that setting the performance goal under the CCI triggered the heightened enforcement. It motivated EPA staff to want to enforce regulations that would contribute to the goal, and to view that enforcement as a high priority. In addition, the comprehensive approach taken under the CCI ensured that the results of the enforcement would noticeably contribute to the goal. This not only provided positive feedback to all parties involved (both within and outside of EPA), but also ensured that individual pollution sources would not be unfairly singled out for enforcement. Finally, the activities undertaken under the CCI generated the monitoring data and information on pollution sources necessary to ensure effective enforcement. Without this information and the feedback it provides, EPA would not have been able to identify and target major pollution sources, and would not have had the type of public support for its enforcement efforts that were created by the increased awareness that resulted from the initiative.

Can EPA continue to rely on threatened or actual enforcement actions as a major source of further improvement in water quality in the river? Water quality has been stalled at or around a grade of B for the last several years. Major improvements resulting from closing CSOs and eliminating illegal sewer hookups have been made and the results are reflected in the grade improvements that occurred over the first five years of the initiative. However, the remaining sources of pollution are likely to be more difficult to control. They appear to be non-point sources that are not subject to regulation. Thus, enhanced enforcement (or the threat of enforcement or imposition of regulation) is not an option for controlling these sources. Controlling these sources will require innovative approaches and proactive behavior by all contributors. The true test of voluntary management-based strategies will be whether, in the absence of regulations, EPA can reduce the remaining sources of pollution sufficiently to meet its fishable/swimmable goal.

Acknowledgments

We acknowledge the useful comments of Cary Coglianese, Jennifer Nash, and participants at the conference "Leveraging the Private Sector: Management-Based Strategies for Improving Environmental Performance," July 31–August 1, 2003, which took place at Resources for the Future in Washington, DC.

Appendix 8-1. List of Enforcement Actions Used to Construct Enforcement Variable

1. 1995: EPA seeks to have comprehensive stormwater management programs in operation in all Lower Charles River communities by December 1996.
2. 1995, October: EPA forms the Charles River Enforcement Team to monitor compliance along the Charles and prosecute violators where necessary.
3. 1996, August 8: Law signed—any project within 200 feet of rivers or streams has to provide compliance reports with local conservation commission. State environmental officials, with backing from more than 100 citizen activist groups, won passage of the Massachusetts Rivers Protection Act.
4. 1996, September 18: EPA orders nine towns, including Brookline, Cambridge, Needham, Newton, and Waltham, to find and remove illegal pipes.
5. 1997, July 16: Under threat of Clean Water Act lawsuits, De-Villars requires public works agencies in 10 communities bordering the river from Dedham to Boston to file plans by the following July 1st, detailing how they will improve street sweeping, catch-basin cleaning, and sewer maintenance, and how they will keep pollution from storm drains.
6. 1997, October 9: In the largest-ever federal environmental penalty against

an educational institution, Boston University agrees to pay a $2 million settlement for oil spills in the Charles River. BU accepted responsibility for two oil spills into the Charles River, one in October 1992, and the other discovered by rowers in January 1996.

7. 1997, October 23: EPA requires the MBTA, the Highway Department, the Metropolitan District Commission, and the Turnpike Authority to identify within 60 days any pipes they own that empty into the Charles and to determine whether they are emitting sewage, oil, or other waste.

8. 1998, March: EPA mails warning letters to 200 major facilities along the river, encouraging them to ensure that they are in compliance with environmental laws.

9. 1998, May 1: EPA makes surprise visits to monitor compliance behavior for the above facilities.

10. 1998, October 16: EPA issues a formal CWA demand letter to the Cambridge Electric Company.

11. 1999: Watertown's Department of Public Works institutes a program in which bikes and a boat will be used to monitor the river.

12. 2000, June 23: EPA strikes an agreement with Watertown, under which the city will pay more than $110,000 to settle claims that it violated federal laws.

13. 2000, July 9: About 170 cities and towns in Massachusetts must comply with new stormwater rules released earlier in the year by EPA.

14. 2002, May 31: EPA fines Boston Sand and Gravel for illegally discharging wastewater over the span of several years into the Charles River basin and two other sites. (This represents one of the largest fines ever imposed for violating CWA.)

15. 2001, April: MIT commits to three environmental projects and pays $150,000 civil penalty to settle violations.

Appendix 8-2. Regression Results from Simultaneous Equation Model

Variables	Model: Log(Bac_Count) $R^2 = 0.476$, N = 336 Coeff.	t-ratio	Variables	Model: Enforcement Actions $R^2 = 0.459$, N = 336 Coeff.	t-ratio
INTERCEPT	3.687*** (0.549)	6.71	INTERCEPT	−1.645** (0.696)	−2.36
STATION 2	0.215 (0.303)	0.71	STATION 2	0.03 (0.443)	0.07
STATION 3	0.645** (.227)	2.01	STATION 3	0.047 (0.456)	0.1
STATION 4	−0.158 (0.322)	−4.9	STATION 4	0.043 (0.443)	0.1
STATION 5	−0.329 (0.307)	−1.07	STATION 5	−0.003 (0.443)	−0.01
STATION 6	0.816*** (0.305)	2.67	STATION 6	0.043 (0.450)	0.09
STATION 7	−1.257*** (0.367)	−3.42	STATION 7	−0.015 (0.443)	−0.03
STATION 8	−1.701*** (.325)	−5.22	STATION 8	−0.134 (0.452)	−0.30
STATION 9	−1.24*** (0.312)	−3.98	STATION 9	−0.222 (0.46)	−0.48
RAINFALL	0.918*** (0.226)	4.07	POLLUTION	−0.1 (0.068)	4.07
ENFORCEMENT ACTIONS	−0.079** (0.031)	−2.59	LAWS	2.763*** (0.171)	−2.59
CSO * RAINFALL	0.1604*** (0.044)	3.61			
DEER ISLAND	0.152 (0.275)	0.55			
STREAMFLOW	0.0301 (0.074)	0.41			
BACT_1	0.349*** (0.0586)	5.96			

** indicates significance at the 5 percent level; *** indicates significance at the 1 percent level. Standard errors are in parentheses.

Notes

1. For annual report card scores, see U.S. EPA (2005). The general leveling out of progress on water quality reflected in the scores since 2000 reiterates our conclusion that further improvements are difficult to achieve.

2. For recent reviews of this literature, see Alberini and Segerson (2002), Khanna (2001), and Segerson and Li (1999).

3. See Metzenbaum (2001) for convincing evidence of EPA staff motivation.

4. Since 1998, EPA's Office of Environmental Measurement and Evaluation (OEME) has implemented a water quality monitoring program based on data from 12 fixed sites along the Charles River. This program is not limited to the grading of water quality based on bacterial counts only and is extended to include other factors like dissolved oxygen, temperature, pH, and suspended solids. However, assignment of the annual grade is still based on the bacterial count only. The focus on bacteria in the grading does not account for any improvement due to reductions in other pollutants, such as oil.

5. For details on the sampling process, see the Charles River Watershed Association Web site at http://www.crwa.org.

6. The water quality data for all monitoring stations are available at http://www.crwa.org/index.html.

7. See Metzenbaum (2001) for a detailed description of the history of the CCI.

8. Participatory actions of the Metropolitan District Commission, as a member of the Task Force 2005, include proper management of the Charles River Dam, restoration of aerators in the Charles River Basin, and control of highway runoff.

9. The Boston Water and Sewer Commission had a program already in place to detect and remove illegal sewage discharges.

10. See the *Clean Charles 2005 Progress Reports* by EPA for the year 1999 and 2000 for further details (U.S. EPA 1999c; 2000b).

11. In 1995, EPA established "Task Force 2005"—The Clean Charles Steering Committee, which comprised state agencies, local communities, and environmental groups, to guide the cleanup project and monitor its success.

12. At the very onset of the project, EPA staff and senior management joined an EPA-sponsored City Year servathon team in stenciling storm drains along the Charles on Beacon Hill and in the Back Bay with the exhortation "Please Don't Dump – Protect the Charles."

13. For details, see the MWRA Web site at http://www.mwra.state.ma.us.

14. See Metzenbaum (2001) for a detailed discussion of this issue.

15. For recent surveys of this literature, see Alberini and Segerson (2002), Khanna (2001), Lyon and Maxwell (2002), OECD (1999), and Segerson and Li (1999).

16. See Alberini and Segerson (2002) for a review of the evidence.

17. Both Pargal and Wheeler (1996) and McClelland and Horowitz (1999) find evidence that community pressure on firms can lead to voluntary reductions in pollution.

18. See http://www.crwa.org.

References

Alberini, Anna, and Kathleen Segerson. 2002. Assessing Voluntary Programs to Improve Environmental Quality. *Environmental and Resource Economics* 22(1–2): 157–84.

Baker, George P. 1992. Incentive Contracts and Performance Measurement. *Journal of Political Economy* 100(3): 598–614.

Besanko, David A. 1987. Performance versus Design Standards in the Regulation of Pollution. *Journal of Public Economics* 34(1): 19–44.

Boston Globe. 1996. Charles River Manages Only a "D" on Its EPA Report Card. April 25, Thursday, City Edition, 95.

Brelis, Matthew. 1995. Brookline Takes Major Step toward Charles Cleanup. *Boston Globe* Jan. 29, City Edition, 32.

Christensen, John. 1982. The Determination of Performance Standards and Participation. *Journal of Accounting Research* 20(2): 589–603.

Courty, Pascal, and Gerald Marschke. 1997. Measuring Government Performance: Lessons from a Federal Job-Training Program. *American Economic Review* 87(2): 383–88.

DEP (Department of Environmental Protection). 2000. *Charles River Watershed 1997/1998 Water Quality Assessment Report.* Commonwealth of Massachusetts, DEP, Division of Watershed Management 72-AC-3. http://www.state.ma.us/dep/brp/wm/wqa/72wqar1.doc (accessed May 20, 2003).

Foran, Michael F., and Don T. Decoster. 1974. An Experimental Study of the Effects of Participation, Authoritarianism, and Feedback on Cognitive Dissonance in a Standard Setting Situation. *Accounting Review* 49(4): 751–63.

Gelastopoulos, Evie. 1997. BU Agrees to Pay $2M for Charles River Spills. *Boston Herald* Oct. 9, All Editions, 016.

Gray, David, Tom Faber, Eric Hall, Mark Voorhees, and Bill Walsh-Rogalski. 2003. Memorandum from David Gray, Tom Faber, Eric Hall, Mark Voorhees and Bill Walsh-Rogalski, EPA, to Robert W. Varney, Regional Administrator, and Ira Leighton, Deputy Regional Administrator, U.S. EPA Region 1. April 29.

Heckman, James, Carolyn Henrich, and Jeffrey Smith. 1997. Assessing the Performance of Performance Standards in Public Bureaucracies. *American Economic Review* 87(2): 389–95.

Holmstrom, Bengt. 1979. Moral Hazard and Observability. *Bell Journal of Economics* 10 (Spring): 74–91.

Howe, Peter J. 1996. EPA Warns Milford of Fines, Suits over Pollution of Charles River. *Boston Globe* September 27, City Edition, B5.

Khanna, Madhu. 2001. Non-mandatory Approaches to Environmental Protection. *Journal of Economic Surveys.* 15(3): 291–324.

Lyon, Thomas P., and John W. Maxwell. 2002. Voluntary Approaches to Environmental Regulation: A Survey. In *Economic Institutions and Environmental Policy,* edited by Maurizio Franzini and Antonio Nicita. Aldershot and Hampshire: Ashgate Publishing.

McClelland, John D., and John K. Horowitz. 1999. The Costs of Water Pollution Regulation in Pulp and Paper Industry. *Land Economics* 75(2): 220–32.

Metzenbaum, Shelley H. 2001. *Measurement that Matters: Cleaning Up the Charles River*. Regulatory Policy Program Working Paper RPP-2001-05. Cambridge, MA: Center for Business and Government, John F. Kennedy School of Government, Harvard University.

MWRA (Massachusetts Water Resources Authority). Undated. Assessing Temporal Changes in Highly Variable Fecal Coliform and Enterococcus Data in Boston Harbor and Its Tributaries by Randomized Block Factorial Anova. http://www.mwra.state.ma.us/harbor/enquad/workshop/arexposter.htm (accessed February 10, 2004).

OECD (Organisation for Economic Co-operation and Development). 1999. *Voluntary Approaches for Environmental Policy: An Assessment*. Paris: OECD.

Pargal, Sheoli, and David Wheeler. 1996. Informal Regulation of Industrial Pollution in Developing Countries: Evidence from Indonesia. *Journal of Political Economy* 104(6): 1314–27.

Segerson, Kathleen, and Na Li. 1999. Voluntary Approaches to Environmental Protection. In *The International Yearbook of Environmental and Resource Economics*, edited by H. Folmer and T. Tietenberg. Cheltenham and Northampton: Edward Elgar, 273–306.

Segerson, Kathleen, and Thomas Miceli. 1998. Voluntary Environmental Agreements: Good or Bad News for Environmental Protection? *Journal of Environmental Economics and Management* 36(2): 109–30.

U.S. EPA (Environmental Protection Agency). 1995. EPA Sets Clean Up Goal for Charles River by Earth Day 2005. *EPA New England Environmental News*, October 22. Press release, available at http://www.epa.gov/docs/region1/charles/pdfs/oldcharm.pdf#search='EPA%20Sets%20Clean%20Up%20Goal%20For%20Charles%20River%20By%20Earth%20Day%202005'.

———. 1997. EPA Issues Annual Charles River Report Card: EPA New England Press Release # 97-7-30. July 15.

———. 1998. Report Card on Charles River Shows Water Quality Improvement : EPA New England Press Release # 98-4-28. April 23.

———. 1999a. EPA Proposes $75,000 Penalty for Wastewater Treatment Violations on Charles River: EPA New England Press Release # 99-4-1. April 1.

———. 1999b. EPA Settles Clean Water Act Case with Medway-Based Pollution Control District; Agreement Benefits Charles River and Franklin Well Field: EPA New England Press Release # 99-8-8. August 13.

———. 1999c. *Clean Charles 2005 Water Quality Report 1998 Core Sampling Program*. Office of Measurement and Evaluation, Region 1.

———. 1999d. EPA Gives Charles River a B-Minus and Announces New Coalition of Private Institutions. EPA 99-4-14. Washington, DC. Press release, available at http://www.epa.gov/newengland/pr/1999/070899a.html.

———. 2000a. EPA gives Charles River a "B" and Announces Innovative "Curtain" at Magazine Beach. EPA Release # 00-04-14. Press release, available at http://www.epa.gov/docs/region01/pr/2000/041400.html.

———. 2000b. *Clean Charles 2005 Water Quality Report 1999 Core Monitoring Program*. December. Office of Measurement and Evaluation, Region 1.

———. 2001. *Charles River Earns Another "B" Grade; Cleanup Halfway to Swimmable Goal*. EPA 01-04-39. Washington, DC: U.S. EPA. Press release, available at http://www.epa.gov/newengland/pr/2001/apr/010439.html.

———. 2005. *Charles River Report Card, May 2005.* http://www.epa.gov/boston/charles2005/evaluate.html

Wald, Matthew L. 1995. Conrail Says It Polluted River and Will Pay $2.5 Million Fine. *New York Times.* July 25, Late Edition – Final, A8.

9

The Role of Management Systems in Stakeholder Partnerships

Andrew A. King

The role of stakeholders in environmental improvement has changed in recent years (Hoffman 1997). In response to diminished expectations that governmental action will solve environmental problems, stakeholders have increasingly turned to forming partnerships with the private sector (U.S. Congress 1994; Schneider 2001). For example, Environmental Defense (formerly the Environmental Defense Fund) and the McDonald's Corporation formed a partnership in 1990 to find a packaging material that would be both less costly and more environmentally benign than existing polystyrene packaging (Crane 1998). Following the enthusiastic public response to this joint project, other partnerships were formed between for-profit businesses and environmental organizations (EOs). In some cases, environmental groups have even formalized these activities by creating dedicated units responsible for initiating and managing partnerships with for-profit firms.

Until recently, perceptions of corporate and stakeholder incentives made such partnerships almost unthinkable (Harrison 1999; Reinhardt 2000). In the traditional view, corporations cause environmental problems when too much of the cost of environmental damage accrues to the public and not enough to the offending firm (Pigou 1962; Barbier 1989). Since such cost transfers are good for a firm's balance sheet but bad for the welfare of surrounding stakeholders, the incentives of managers and stakeholders seem irreparably misaligned. Indeed, stakeholder groups have viewed the return of these costs to the firm's balance sheet as a central part of their mission. To this end, stakeholder groups have lobbied government for stricter regulation; sued firms in civil courts; and organized protests, boycotts, and other forms of community pressure (U.S. Congress 1994). Not surprisingly, relations between managers and stakeholders have often been antagonistic (Harrison 1999).

In recent years, however, two perceptual changes with respect to the cause of environmental problems have eased tensions between stakeholders and managers. First, a significant portion of environmental problems is now thought to arise from managerial and organizational inefficiency (Hart 1995; Hart and Ahuja 1996). Notions of managers and organizations as efficient optimizers have been replaced by the view that most firms operate far from an efficient frontier, and that such inefficiency causes both extraneous harm to society and additional cost to these firms (Porter 1991; Ashford 1993). Both stakeholders and managers share incentives to identify and reduce these inefficiencies. As a result, stakeholders have taken on a more consultative role in order to help firms identify such win-win opportunities.

Increased awareness of the difficulties in satisfying a growing demand for green products represents an equally important second change in perception. Green products are particularly hard to market and sell because their "greenness" is not easily verified by the consumer (Reinhardt 2000). Although a consumer can often assess a product's quality through inspection, the environmental attributes of a product are usually impossible to determine. For example, a consumer can inspect and smell coffee beans to determine their quality, but cannot determine whether the coffee has been grown by a farm that preserved the forest canopy or by one that destroyed it. The inability to verify a company's environmental claims makes consumers less willing to pay for claimed environmental performance (Reinhardt 2000). Solving this information problem is in the interest of both stakeholders and corporations.

Although partnerships to reduce inefficiencies or to serve green markets have been extensively and positively reported in the popular press, they have so far received only preliminary attention from scholars (Milne et al. 1996; Hartman and Stafford 1997; Crane 1998; Stafford et al. 2000; Rondinelli and London 2003). I build on this preliminary work by considering how such partnerships manage one of the most important elements of any technological development or discovery process: the protection of intellectual property. Specifically, I show that a key impediment to forming partnerships is the risk of disclosure of a corporation's *intellectual property*, a term I use broadly to refer to any confidential business information the disclosure of which would adversely affect a firm's economic interests. EOs have the incentive to diffuse information about new technologies and business practices throughout industry, while firms have an interest in keeping information about new win-win technologies and practices to themselves. These diverging incentives over intellectual property limit opportunities for partnerships, influence partnerships' goals and structures, and ultimately increase the importance of environmental management systems.

Environmental management systems, I conclude, can play a critical role in overcoming the informational impediment to successful partnerships between environmental organizations and corporations. By structuring partnerships to focus on management systems, corporations can protect their business information while gaining value through the engagement of environmental organizations. Because management systems provide information about a firm's broader goals and its plans for meeting them, rather than technical information about business operations, they provide a mechanism through which environmental organizations may be able to observe and shape corporate behavior without needing access to sensitive business information.

Intellectual Property Protection as a Barrier to EO/Corporate Partnerships

Alliances and partnerships among for-profit firms arise for many reasons: firms may wish to harness unique resources, reduce individual investment risk, or develop publicly valuable technology (Das and Teng 2000). In some cases, these partnerships and alliances generate important intellectual property. Academic studies suggest that for partnerships to be successful, two central governance problems must be solved (Das and Teng 1999). Systems must exist to regulate the contribution and reward of all partners, not just some; and systems must exist to protect and distribute any intellectual property that is developed.

For-profit corporations generally benefit either by maintaining property rights to any technology that they develop or by keeping such technology secret. When for-profit firms form a technology-development alliance, each firm can reasonably assume that the others will want to protect any intellectual property that is generated. Individual firms may differ on how to distribute the costs of developing this intellectual property or how to allocate the returns from its use, but they are all likely to share a desire to maximize the alliance's private return by protecting the property from transfer to third parties.

Partnerships that involve EOs face added difficulties in governing intellectual property because EOs and for-profit corporations benefit differently from new environmental technology. EOs usually do not share in the private return of any new technology or technique developed through the partnership. Instead, they benefit from the public benefits provided by the use of a new technology. Thus, unlike corporations, EOs often have strong incentives to disclose any intellectual property that is developed through a partnership.

It might seem that these conflicting objectives about the disclosure of intellectual property could be solved through the use of a contract that required or induced an EO partner to keep any developed intellectual property private. For example, the EO might share in any profits gained from new technology. Such contracts are often not feasible, however, because such restrictions, payments, or licenses could create conflicts of interest for environmental stakeholders. To the extent that EOs rely on support from this broader group of stakeholders interested in environmental protection, they will be unwilling to accept such direct payments, because to do so would imply that the EO is no longer solely dedicated to furthering stakeholder interests. For this reason, EOs typically refuse to accept payments from alliance partners and require that partners agree to make public all information generated by the alliance (Laurent 2003).

Corporations can respond in several ways to the added risk posed by partnering with EOs. They can choose to avoid such partnerships and instead form alliances with for-profit firms with similar skills. Alternatively, they can focus partnerships with EOs on subjects that are far from their existing intellectual property. Finally, they can structure partnerships so that organizational attributes make it difficult to transfer intellectual property. For example, they may embed intellectual property in tacit practices and routines, or make its value contingent on a unique aspect of the firm.

Although partnerships with EOs increase intellectual property risk, they have at least one distinct advantage. EOs' reliance on stakeholder contributions increases their credibility with environmental stakeholders. Thus, when environmental impact is a major part of a technology-development effort and it is difficult to communicate the effort's environmental attributes credibly, EOs can act as important conduits to stakeholders. EOs can credibly communicate information about attributes that their stakeholders cannot observe. For example, EOs might be able to certify the use of a certain technology or environmental practice that they observed through their interactions with the corporation.

The history of some important partnerships between EOs and corporations reveals how intellectual property issues shape the structure and role of the partnerships. In particular this history reveals the important role played by environmental management systems. In the sections below, I briefly outline the history of several partnerships. I begin with one that failed to provide a sustained benefit to both partners. Although it provided tremendous value to the partnering EO and to environmental stakeholders, it did not provide the corporate partner with a long-term economic benefit. I then discuss three examples of partnerships between corporations and Environmental Defense, the environmental organization that has most strongly championed such partnerships. These examples reveal how partnerships can be structured to protect intellectual property and to provide mutual gain to both partners.

Partnership Problems: The Greenfreeze/DKK Partnership

The collaborative effort of DKK Scharfenstein and the environmental organization Greenpeace to develop a completely CFC-free refrigerator (the "Greenfreeze" partnership) is one of the most famous examples of a partnership between an EO and a for-profit firm. Yet it must be judged to be a mixed success. Although it provided tremendous benefit to the environment, it failed to provide a sustained benefit to the corporate sponsor.

The principal environmental harm from refrigerators comes from the use of chlorofluorocarbons (CFCs) in the production of insulation and as a coolant in the refrigerator itself. When released into the atmosphere, CFCs cause dramatic catalytic damage to the ozone layer. Scientific recognition of the damage to the ozone layer and the Montreal Protocol limitation on the production of CFCs forced companies to seek alternative blowing agents and coolants for refrigerators. DuPont, the principle producer of CFCs, invented a series of less-harmful products called hydrochlorofluorocarbons (HCFCs), but these products continued to pose some environmental risk and also created some technical problems.

CFCs and HCFCs are not the only materials that can be used as coolants. Other materials, chiefly hydrocarbons such as propane and butane, have been used as coolants since the earliest days of refrigeration, but in most applications these materials had been superseded by CFCs. In the early 1990s, both Greenpeace and the Dortmund Institute of Hygiene renewed the search for a viable hydrocarbon-based refrigerator technology (Stafford and Hartman 2001). Based on his research, Wolfgang Lobeck, head of the Atmosphere Campaign at Greenpeace Germany, became convinced that a hydrocarbon process could match or exceed the performance of CFC-based refrigerators, and he began to approach German producers with the idea. Rebuffed by mainstream German producers, he eventually found one interested producer—DKK Scharfenstein, an East German company that was finding itself unable to compete in the newly unified Germany (Stafford and Hartman 2001). DKK's sales and workforce had fallen by 90 percent following reunification, and it was threatened with closure (van der Linde 1994). DKK did hold, however, technical leadership in the use of hydrocarbons rather than CFCs or HCFCs as blowing agents in making foam insulation (van der Linde 1994). Because its location in East Germany had restricted access to CFCs, the company had years of experience with a process that used inexpensive propane and butane. If it could combine this blowing technology with a hydrocarbon-based coolant technology, DKK would be able produce a refrigerator that was totally free of CFCs.

Greenpeace and DKK formed a partnership to make a CFC-free refrigerator a practical marketable product. In 1992, DKK agreed to make 10 prototypes of the hydrocarbon refrigerator to demonstrate the feasibility of the design for a small fee of DM 24,000 (van der Linde 1994). Greenpeace's contributions were its research on the use of hydrocarbons as a coolant and its creation of demand for the new product. Greenpeace certified the product as CFC-free, and—at a cost of only US$68,000—signed up 70,000 purchase orders before production had even begun (van der Linde 1994).

Despite the apparent popular interest in a CFC-free refrigerator, the rest of the German industry discounted the market and technical value of hydrocarbon refrigerators. They argued that such refrigerators would use energy inefficiently, that the foam insulation would have reduced insulating power, that the propane/butane hydrocarbon mixture would pose a significant explosion risk, and that the use of such a coolant would prohibit the creation of a "frost-free" refrigerator. Sharing this pessimism, the German privatization agency Treuhandanstalt announced its intention to close DKK (van der Linde 1994).

DKK quickly discovered solutions to each of these potential problems. In fact, with some modifications, the hydrocarbon refrigerator proved to have several advantages over CFC/HCFC refrigerators. Hydrocarbons were much less expensive to purchase than HCFCs, and they did not chemically react with inexpensive lubricants (as did HCFCs). This made it possible to retrofit existing refrigerators and to use existing designs and lubricating materials. DKK also demonstrated that hydrocarbon refrigerators actually allowed higher energy efficiency. Safety problems also presented few barriers. DKK demonstrated that the refrigerators used less propane and butane than a common cigarette lighter, and in December 1992, the product was tested and judged to be safe by the rigorous Bavarian testing agency Technisher Uberwachungsverein. The product was nominated for a prestigious Blue Angel environmental award. Encouraged by the technical successes and early sales, a group of investors purchased the firm and changed its name to Foron Hausgerate.

After the launch of the "Clean Cool" product in February of 1993, Greenpeace turned its attention to other producers of refrigerators. Through the use of public advertising and private negotiations, it encouraged these producers to manufacture CFC-free refrigerators. It also transferred information to potential new producers. John Harris of Calor Gas commented, "We were surprised by the amount of scientific research, technical data, and contacts Greenpeace had amassed. . . . [It] would have taken a good deal of time and money to establish contacts like this" (Stafford and Hartman 2001). As a result of the apparent demand for the product and increased information about its feasibility, all the German producers quickly announced that they would also design and produce a CFC-free hydrocarbon refrigerator.

Hydrocarbon refrigerators designed to compete with the DKK/Foron refrigerators arrived on the market by early 1994. Despite an infusion of DM 35 million in capital into Foron, the company lost market share to competitors (van der Linde 1994). In the midst of increased competition, and lacking any protectable intellectual property, the company's profits vanished. By June 1994, Anna Tomforde of *The Guardian* reported that "Foron GmbH, pioneer of the world's first 'green fridge,' has run into difficulties because major electrical producers have copied its success" (Tomforde 1994). Siegfried Schlottig, Foron's spokesman, noted that "[t]he market has responded and we are left behind. We did not think it would happen so quickly" (Tomforde 1994).

As a result of the Greenpeace/DKK partnership, hydrocarbon refrigeration technology was rejuvenated and eventually came to be a major new product line. Today hydrocarbon refrigerators dominate the European market and represent about 50 percent of refrigerator production in China. Yet the technology and market pioneer, DKK/Foron, was quickly displaced. The company had been weakened by years of mismanagement, and even with an infusion of capital it is possible that it had little chance of survival in the competitive German market. Nevertheless, Greenpeace's efforts to diffuse the technology to competitors undercut DKK/Foron's ability to reap gains from its innovative technology. Lacking control over its intellectual property, DKK/Foron was forced to compete on cost and service—two areas where its larger competitors had substantial advantages.

Mechanisms for Solving Partnership Problems: The Cases of Environmental Defense

Three recent partnerships in the United States between several corporations and Environmental Defense (ED) illustrate ways that partnerships can be structured to protect intellectual property. These partnerships include mechanisms for appropriating some return from developed intellectual property.

The Federal Express/ED Partnership

In the Spring of 2000, ED approached Federal Express with a proposal to create a more efficient standard postal delivery vehicle, the Class 4 Walkin. Fred Smith, the CEO of Federal Express, enthusiastically endorsed the idea. As part of the joint effort, the partners created a team to develop a challenging set of specifications for a new vehicle. Eventually, these called for a vehicle with 50 percent better fuel efficiency, 90 percent fewer particulate emissions, and 70 percent reduced output of NO_x and SO_x (Laurent 2003). The specifications were then written into a standard industry Request for Information. More than 50 potential supplier companies responded with proposals, all of which suggested the use of a diesel hybrid electric engine (Steffan 2003). The team then selected four finalists from the 50 proposals, and these companies were asked to create test vehicles. Eventually, Eaton Corporation's vehicle was selected for in-use testing (Steffan 2003). Eaton is now in the process of producing 10 vehicles for field testing.

In the partnership, Federal Express took the lead role in creating the specifications for the new truck, writing the Request for Information and the Request for Proposals, evaluating the responses, and testing the vehicles. According to participants, ED's primary role was to act as a professional "facilitator" and to make sure that environmental issues received fair treatment (Laurent 2003). ED staff also helped to "get the project press" (Laurent 2003).

The benefits of the project to ED's stakeholders are clear. If Federal Express eventually replaces all of its 10,000 Class 4 Walkin vehicles, the new vehicles will reduce smog-causing emissions by 75 percent. The competitive advantages to Federal Express are less apparent: Federal Express bears part of the cost of the development effort, yet their competitors will also be able to use the new vehicles. "Nothing prevents UPS from buying the vehicles too," commented one of Federal Express's managers on the project (Steffan 2003). Alternatively, other firms may also be able to learn about the project from public information and shorten the development time of competing vehicles. Indeed, UPS is already working with Navistar International to develop a competing hybrid diesel-electric vehicle.

Although personnel at Federal Express believe that the project will provide limited competitive advantage, they do point to some possible benefits. First, Federal Express may be able to appropriate some return from the project by matching the truck design with specific Federal Express systems. Federal Express's role in the design process allowed them to tailor the control system for the Eaton vehicle for loadings and usage common to Federal Express vehicles (Steffan 2003). As a result, Federal Express expects to retain a slight advantage in fuel economy, even if other companies also purchase the vehicles. Second, Federal Express may be able to appropriate some return through a slight first-mover advantage. Because of their work with Eaton on the development of the prototype, they have the right of first refusal on production vehicles (Steffan 2003). Finally, Federal Express may be compensated for its development investment. ED is currently seeking funds to offset some of the cost of the preproduction and production vehicles (Laurent 2003; Steffan 2003).

The SC Johnson/ED Partnership

The SC Johnson/ED joint task force grew out of interaction between Sam Johnson, chairman of SC Johnson, and Fred Krupp, executive director of ED, on the President's Council on Sustainable Development (Alston 2003). The task force was charged with integrating "environmental considerations into the concept, design, development, and review of SC Johnson's products and packaging" (Alston and Roberts 1999). The project had two main elements: (1) the creation of a system for bringing environmental considerations into the early stages of product conceptualization and design, and (2) the formation of a better understanding of consumers' preferences regarding the environment (Alston and Roberts 1999).

Before the project began, SC Johnson had already developed a set of systems for encouraging product designers to consider environmental issues. Ken Alston had developed a computer-based tool that helped designers understand the environmental impact of different design choices (Alston 2003). Working with ED added richness and detail to the existing criteria in Alston's system. It also made the analysis more complex and potentially more accurate. The team struggled to make the system simple to use and understand while legitimizing it with environmental stakeholders (Alston 2003). According to participants, working with ED "made [the system] more robust . . . people outside could see that it went beyond generalities" (Alston 2003).

Unfortunately, the system "fell into disuse almost from the get-go" (Alston 2003). The added complexity of the system made it harder for designers to understand how it worked. They quickly began to doubt and distrust the system. Despite this problem, know-how that was developed on the project "still got used" (Alston 2003). In particular, what the company learned through the project helped it improve its materials selection program.

After the end of the project, the computer software moved to ED, but according to engineers at SC Johnson, it was unlikely to have much use for either ED or other corporations. "I could never understand how they expected it to work," Alston said (Alston 2003). He argued that the system only worked within SC Johnson because they had carefully collected important data inputs to the system. These data were "never part of the collaboration" and without them, the software tool was nearly useless (Alston 2003).

The Starbucks/ED Partnership

In August 1996, Starbucks and ED formed a partnership to "reduce the environmental impacts of serving coffee in Starbucks retail stores" (Starbucks/ED 2000). This partnership explored the possibility of improved packaging for hot beverages, a concern that arose because some consumers tended to "double cup" in order to further insulate hot beverages (Starbucks/ED 2000).

The project also confirmed the value of using ceramic cups rather than paper ones for in-store consumption. The team performed a life-cycle analysis of the paper cups to determine their true cost. It then compared this cost with the cost of purchasing ceramic cups. It also performed in-store tests of consumer preferences and evaluated ceramic cup use at several target stores (Starbucks/ED 2000).

In designing a new, more insulated cup, the team created several alternative cup designs and tested them with customers. Eventually, the team selected a cup that used embossing to create an insulating thermal layer. Unfortunately, the cup proved difficult to produce, and early prototypes had a tendency to "wick" coffee onto the fingers of customers (Starbucks/ED 2000). Eventually, upper management at Starbucks chose to not implement the new cup design because of remaining concerns about the cup and because increasing use of insulating sleeves was reducing the double-cupping problem (Starbucks/ED 2000).

Methods for Protecting Intellectual Property in EO/Corporate Partnerships

These three partnerships with ED reveal various ways that corporate partners can protect their interests when partnering with EOs. For example, Federal Express retained a right of first refusal on new vehicles built by Eaton Corporation, SC Johnson retained control over the data in its product design system, and Starbucks reserved its ability to decline to proceed with any alternative cup developed through the partnership. Still more generally, partnerships can take advantage of at least three strategies for protecting intellectual property:

- selecting projects to reduce IP leakage,
- constructing barriers to information transfer, and
- reducing the risks by certifying environmental attributes or effort.

Selecting Projects

The easiest way for a firm to protect its intellectual property is to refuse to form a technological development partnership with an EO. Alternatively, firms can choose to select projects for partnerships that address issues that are not central to the firm's competitive advantage. In two of the ED partnerships addressed earlier, the projects focused on issues far from the main intellectual property of the firm. Although Starbucks uses coffee cups and Federal Express uses delivery trucks, the actual design of these elements is not a basis of competitive advantage for either firm. As a result, there was little risk that the alliance could reveal and damage intellectual property central to core capabilities.

Similar strategies for intellectual property protection have been followed in other partnerships. A Citigroup/ED alliance focused on the use of recycled paper and double-sided copying in Citigroup. Clearly, recycling and paper consumption are peripheral issues for the firm and are far removed from the corporation's core skills in banking and finance. Likewise, a United Parcel Group (UPS) alliance with ED increased the use of recycled material in boxes and did not address issues related to UPS's competitive advantages in scheduling, materials handling, and tracking.

Constructing Barriers to Information Transfer

ED partnerships have contained elements that make it difficult to transfer any intellectual property that emerges as a result of the partnership. Information can be protected by explicit rights, or it can be protected by making it hard to transfer. No one holds a property right on how to surf or ride a bike, but it is still difficult to transfer such learning. When knowledge is hard to describe and communicate, like the information needed to ride a bike, it is called "tacit knowledge" (Hayek 1978). Two of the ED partnerships discussed involved such tacit knowledge, which may have protected this information.

The SC Johnson partnership utilized a nonproprietary computer system that encouraged designers to consider environmental issues in product and packaging design, but they embedded this system in a unique managerial and technical context. Previous research has shown that design learning cannot be effectively transferred by simply moving computerized engineering design tools from one location to another. Such systems are notoriously difficult to transfer because the issues involved in using the system remain tacit (Zuboff 1984). For example, information about the stage in which the SC Johnson system would be used, the organization of users, and their role in project teams was embedded within relatively opaque and unique routines and management systems.

Federal Express may have used an approach different from SC Johnson's to appropriate some return on the intellectual property developed as part of its vehicle development partnership. Although it retained no exclusive right to buy vehicles from Eaton, Federal Express may have been able to acquire some competitive advantage by tailoring the design of the hybrid trucks to match the Federal Express system. According to managers at Federal Express, one of the critical elements of the truck design was the computer system that managed the use of the diesel engine and the electric motor. Considerable expense went into developing this system, and it greatly affected emissions and fuel efficiency. By influencing the design process, Federal Express managers hoped to gain an efficiency advantage over other uses.

The Federal Express story also suggests another strategy for intellectual property protection: the transfer of intellectual property rights and the investment of a third party. In the case of the Federal Express partnership, most of the development cost and intellectual property were retained by the supplier. Federal Express's investment included the initial specification of the vehicle's performance, the evaluations of the responses to the Request for Information, and the testing of prototype vehicles from two finalist companies. Although significant, many of these expenses would have been required for sourcing any new vehicle—regardless of its type. Federal Express's only commitment was to purchase 20 vehicles, and even these costs may be offset: representatives of FedEx expect that ED will find a sponsor willing to cover some of this cost.

Reducing Risks by Certifying Environmental Attributes or Effort

Finally, the ED examples illustrate how partnerships can reduce the risk of a project by certifying the environmental attributes of its outcome or the good faith effort that was exerted to find a win-win solution.

A central problem in any technical development partnership is uncertainty about its eventual outcome. The more uncertain the outcome, the more difficult it is to justify an investment. For environmental projects, this problem is complicated by the intangibility of environmental issues and the unpredictability of environmental stakeholders. These stakeholders may not believe that a product has environmental attributes, or they may doubt that these attributes represent a clear environmental gain. In the celebrated partnership between ED and the McDonald's Corporation, ED helped McDonald's manage the risks associated with shifting away from its old clamshell packaging. When two Harvard professors later criticized the McDonald's switch from clamshell containers to paper wrappers, claiming that the problems with polystyrene had been overstated and the feasibility of recycling had gone incompletely analyzed (Rayport and Lodge 1990), ED effectively defended McDonalds against these accusations. ED could present an effective rebuttal because they had been involved in the management of the project and so knew precisely the diligence of the effort and the tradeoffs inherent in different designs.

Credible communication of environmental attributes is a critical element of any development partnership. To communicate such attributes, stakeholders can provide information about the output of the partnership, or the inputs used in the partnership (or both). Providing information about the inputs used reduces the technological risk to the corporation. Inputs can be agreed upon before the partnership is formed, and their cost can be determined with considerable certainty. For this reason, many technological development contracts employ an input-based, cost-plus payment system. Companies are rewarded based on the cost of their inputs, plus some agreed-upon profit. In a similar way, measurement and certification of the input of corporate partners with EOs allows corporations to reduce their risk. At least the corporations know that they will be certified as having put in a good effort, even if it doesn't lead to improvements.

In two of the three ED examples, the EO certified the management process of the project, but not the output of the effort. The Starbucks project did not result in the use of a different hot cup, yet ED reported that the project was a success because the team had conducted an investigation of the financial and environmental implications of alternative designs. In the SC Johnson case, ED could only document changes to the design process, but could not opine whether these changes influenced the environmental attributes of developed products. When asked what SC Johnson got out of the partnership with ED, one manager replied, "third-party endorsement" (Alston 2003). ED documented the effort to create an effective environmental design tool. It did not document, however, the degree to which the tool was used or the degree to which it resulted in improved environmental design.

In the Federal Express partnership, it was possible to certify the eventual output of the design (emissions, fuel efficiency, and so on), but even here the ability to provide information about the design process provided a strategic advantage. Even if consumers know the technical specifications of the selected truck, they might be uncertain whether this design should receive their approbation because they do not know what other options have been considered and rejected. They may rationally assume that the design benefits the company, but they have no information about how the company chose between provision of public goods and private gain. By being directly involved in the management system of design, ED made it credible that public benefits had been given full weight during the project. As one ED representative said, his role was to "make sure Federal Express stayed committed to the environmental goals" (Laurent 2003).

Thus, all three ED projects involved certification by ED of management inputs into technological improvements and thereby reduced the risk of failure to the partnership. If unforeseen technical problems were uncovered (as in the Starbucks case), ED would certify that the failure was not due to a lack of effort on the corporation's part. Such certification was a critical element of every project. When asked what ED's major contribution was to the project, one corporate representative thought for a moment and then responded: "credibility" (Steffan 2003).

It is also possible that certification of management processes provides a larger benefit to organizations by providing credible evidence that the organization is more environmentally responsible. Although there is no clear evidence that the corporations involved in these partnerships were more environmentally benign than competing firms, it is possible that stakeholders interpret their participation as such. Indeed, there is some evidence that corporate managers believe this to be the case. For example, UPS's participation in an early partnership with ED was one of the motivations for the Federal Express project. Federal Express did not want to be seen as the less environmentally responsible carrier. Thus it may be that participation in such partnerships provides some direct value to a firm's image. Since a firm's image or reputation cannot easily be transferred, this may allow the firm to gain a proprietary advantage from the project.

Table 9-1 summarizes the main elements of all four partnership examples. All the ED projects used certification of the management process to provide value to corporate partners. All had systems for protecting existing intellectual property. Two had systems for appropriating some return from any significant investment in new intellectual property. The one exception, Starbucks, failed to deliver any significant new technology or technological change.

It is interesting to note that the Greenfreeze partnership differs in almost every structural attribute. Unlike the ED partnerships, the Greenfreeze development effort related directly to the company's core product, thereby increasing the risk of intellectual property transfer. The development was not off-loaded to a supplier, but borne directly by the firm. No mechanism for intellectual property protection was created by DKK/Foron, nor did the project certification focus on the process of development and change; rather it concerned the

TABLE 9-1. Attributes of EO/Corporation Partnership Examples

	Starbucks	*Federal Express*	*SC Johnson*	*Greenfreeze*
Project Selection	noncore; little investment	noncore; supplier bears risk	core process	core product
Increasing the Difficulty of Information Transfer	none	contingent upon product design	tacit process; core element kept private	none
Risk Reduction through Management System Verification	yes	yes	yes	no

actual output. As a result, with Greenpeace's assistance, it was relatively easy for other producers to acquire the technology needed for a CFC-free refrigerator and to compete with DKK/Foron.

The Role of Environmental Management Systems

The case examples discussed in this chapter reveal that management systems are a critical element of EO/corporate alliances. Management systems provide a visible investment in time and effort that can be observed and certified by an EO. To the extent that they are difficult to change quickly, management systems help reduce the propensity of corporations to shirk from a project during or after a period of partnership with an EO. They also facilitate the protection of intellectual property and may be useful in ensuring that partnerships provide a viable and credible signal of superior environmental quality.

Given the uncertain process of technological development, partners cannot easily gauge the extent to which each party is actively working for the partnership's goals. One way to overcome this is to observe input into development processes. Perhaps the most critical input is the management process used by the corporate partner. This process determines the resources allocated to the project, the kinds of options considered, and the criteria used for selecting outcome. Because this process is relatively observable, it provides a mechanism for EOs to monitor corporate effort. For example, EOs can observe the alternatives considered in formulating a new product or the practices used in sourcing raw materials. As in the case of SC Johnson, corporations sometimes routinize the system to aid observation of effort. Computer programs such as SC Johnson's STEP (Success through Environmental Progress), or formal procedures such as Chevron's DEMA (for Decision Making) allow both managers and EO alliance partners to observe otherwise unobservable activities (Reinhardt 2000).

Management systems also provide a vehicle for preventing moral hazard after partnerships have ended. Management systems and processes are notori-

ously difficult to change. Once established and made into a routine, they tend
to take on a life of their own (Coglianese and Nash 2001). Thus, ED may be able
to assume that once a corporation has established an effective management
system, this system will continue to operate after the project has ended.
McDonald's, for example, seems to have established a routinized internal
process for environmental improvement.

Finally, a focus on environmental management systems may help to restrict
the types of firms that participate in EO partnerships, and thereby give the
partnerships some real meaning. For partnerships to provide credible information to stakeholders, they must be inherently more costly to firms with poor
environmental performance. If this is not the case, "adverse selection" may
result, wherein firms with poor performance rush to form such partnerships in
an attempt to "greenwash" their corporate image. Firms with poor performance
tend to have less intellectual property, and thus have less to fear from a partnership with an EO. They also tend to have more to gain from any intellectual
property transferred to them, making such partnerships more desirable.

Environmental management systems may provide the most significant
counterbalance to such adverse selection by increasing the cost to form a partnership with an EO for firms with poor environmental performance. Since the
management systems and practices of partnering firms will be observed by EO
partners, firms with poor performance (and thus poor management systems)
may fear that EOs will observe and report such practices. Since the risk of such
exposure is greater for firms with poor performance, it may help offset the tendency for adverse selection. Stakeholders may be able to assume that the very
existence of a partnership provides evidence that the corporate partner has
green management. The McDonald's/ED experience seems to bear this out.
Although several companies copied the McDonald's switch from polystyrene to
paper wrappings, it is McDonald's that continues to gain the approval of environmental stakeholders. As one SC Johnson official said about his company's
partnership with ED, "[t]he critical thing is that we got a company endorsement, not a product endorsement" (Alston 2003).

Conclusion

In recent years, we have seen increased understanding that EOs and corporations can engage in constructive partnerships. Whatever the merits of such
efforts, these partnerships face difficult barriers. Although EOs and corporations can both benefit from any environmentally friendly technology that
emerges from such partnerships, they benefit in different ways. Corporations
benefit from keeping their technology private, but EOs benefit from making it
public. As a result, without careful structuring, partnering with EOs can
increase the risk of damaging a company's intellectual property and limiting its

ability to appropriate returns on any intellectual property developed in the partnership. Partnerships therefore need to be designed to reduce the risk of harming a company's business interests and to protect any intellectual property that is developed. To maintain both the credibility of a company's green efforts as well as to protect its intellectual property, management systems play a critical role.

Several implications for environmental policy follow from this analysis. First, problems with protecting intellectual property are likely to impede widespread use of technology-development partnerships. Those that *are* formed are likely to address elements peripheral to the organization. They are likely to consider efforts that lead to the creation of organizational procedures—and efforts in which the procedure is both important to the final outcome and can be observed by the stakeholder.

Second, this analysis suggests that alliances to certify sourcing will continue to be the most common form of partnership. In these partnerships, both parties develop rules for sourcing inputs into the corporation (e.g., recycled paper, hormone-free chicken, low-emission vehicles), and then the EO certifies that these processes are in use. Such partnerships minimize the investment to the corporate partner, protect the intellectual property of suppliers, and allow direct certification of a management process (the sourcing system).

Finally, this analysis suggests that the most significant technological breakthroughs from partnerships may, ironically, come from peripheral and possibly lagging companies in an industry. Such corporations have less to fear from intellectual property transfer and more to gain from the creation of a radical new technology. In the case of Greenfreeze, managers at DKK/Foron were willing to throw caution to the wind and join a partnership regardless of the difficulties with intellectual property protection. They recognized that their company did not have the capability necessary to compete with existing CFC/HCFC refrigerators. They did, however, hold a narrow lead in the use of hydrocarbons, and—facing imminent closure—they had little to lose. Greenpeace hoped to demonstrate that a hydrocarbon refrigerator was possible and thereby destroy the existing CFC/HCFC design. With nothing to lose, the two joined forces to change the business game. As Wolfgang Lobeck himself noted, "It was a piece of luck that this firm [DKK/Foron] was there, that it was up to its neck in troubles, that . . . it] could develop the propane/butane prototype on its own. It was a piece of luck that we could win one company over to our way of thinking, and that this firm could turn facts quickly into marketable realities" (Stafford and Hartman 2001).

Perhaps Greenpeace was indeed lucky that a peripheral player such as DKK/Foron existed in the refrigerator industry, but it was not luck that DKK/Foron was the one company willing to cooperate with Greenpeace. The analysis presented in this chapter suggests that it is just such peripheral players that will be willing to overlook the risk inherent in partnering with an environmental organization to develop novel technology. Unlike more central firms that have a strong incentive to protect their intellectual property through the means discussed in this chapter, these more peripheral organizations will attempt creative technological leaps of faith. Although they may fail in so doing, their actions can provide important technology for meeting challenging environmental problems.

References

Alston, Ken. 2003. Personal communication between Ken Alston, director of Education for GreenBlue, previously director of Sustainable Product Innovation at SC Johnson, and the author, October 17.

Alston, Ken, and Jackie Prince Roberts. 1999. Partners in New Product Development: SC Johnson and the Alliance for Environmental Innovation. *Corporate Environmental Strategy* 6 (2): 110–28.

Ashford, Nicholas A. 1993. Understanding Technological Responses of Industrial Firms to Environmental Problems: Implications for Government Policy. In *Environmental Strategies for Industry*, edited by Kurt Fisher and Johan Schot. Washington, DC: Island Press, 277–307.

Barbier, Edward B. 1989. *Economics, Natural-Resource Scarcity and Development*. London: Earthscan.

Coglianese, Cary, and Jennifer Nash. 2001. Environmental Management Systems and the New Policy Agenda. In *Regulating from the Inside: Can Environmental Management Systems Achieve Policy Goals?* edited by Cary Coglianese and Jennifer Nash. Washington, DC: Resources for the Future, 1–25.

Crane, Andrew. 1998. Exploring Green Alliances. *Journal of Marketing Management* 14(6): 559–79.

Das, T. K., and B. S. Teng. 1999. Managing Risks in Strategic Alliances. *The Academy of Management Executive* 13(4): 50–62.

———. 2000. A Resource-Based Theory of Strategic Alliances. *Journal of Management* 26(1): 31–61.

Harrison, Kathryn. 1999. Talking with the Donkey: Cooperative Approaches to Environmental Protection. *Journal of Industrial Ecology* 2(3): 51–72.

Hart, Stuart. 1995. A Natural-Resource Based View of the Firm. *Academy of Management Review* 20(4): 985–1014.

Hart, Stuart, and Gautam Ahuja. 1996. Does It Pay to Be Green? An Empirical Examination of the Relationship between Emissions Reduction and Firm Performance. *Business Strategy and the Environment* 5(1): 30–7.

Hartman, Cathy L., and Edwin R. Stafford. 1997. Green Alliances: Building New Business with Environmental Groups. *Long Range Planning* 30(2): 184–96.

Hayek, Friedrich A. 1978. *New Studies in Philosophy, Politics and Economics*. Chicago: University of Chicago Press.

Hoffman, Andrew J. 1997. *From Heresy to Dogma: An Institutional History of Corporate Environmentalism*. San Francisco: The New Lexington Press.

Laurent, Chad. 2003. Personal communication between Chad Laurent, research associate, Alliance for Environmental Innovation, and the author. October 16.

Milne, George R., Easwar S. Iyer, and Sara Gooding Williams. 1996. Environmental Organization Alliance Relationships Within and Across Nonprofit, Business, and Government Sectors. *Journal of Public Policy and Marketing* 15(2): 203–15.

Pigou, A. C. 1962. *The Economics of Welfare*. London: Macmillan.

Porter, Michael. 1991. America's Green Strategy. *Scientific American* 264(4): 168.

Rayport, Jeffrey F., and George C. Lodge. 1990. The Perils of Going "Green." *St. Petersburg Times*, December 9, 7D.

Reinhardt, Forest L. 2000. *Down to Earth: Applying Business Principles to Environmental Management*. Boston, MA: Harvard Business School Press.

Rondinelli, Dennis, and Ted London. 2003. How Corporations and Environmental Groups Cooperate: Assessing Cross-Sector Alliances and Collaborations. *Academy of Management Executive* 17(1): 61–75.

Schneider, A. 2001. More "Green" Partnerships Sprouting. *Kiplinger Business Forecasts.* August 2.

Stafford, Edwin, and Cathy Hartman. 2001. Greenpeaces's Greenfreeze Campaign. In *Ahead of the Curve: Cases of Innovation in Environmental Management*, edited by Ken Green, Peter Groenewegen, and Peter S. Hofman. Boston: Kluwer Academic Publishers, 107–31.

Stafford, Edwin R., Michael Jay Polonsky, and Cathy R. Hartman. 2000. Environmental NGO-Business Collaboration and Strategic Bridging: A Case Analysis of Greenpeace-Foron Alliance. *Business Strategy and the Environment* 9(2): 122–35.

Starbucks/ED. 2000. Report of the Starbucks Coffee Company/Alliance for Environmental Innovation Joint Task Force. Washington, DC: Environmental Defense. http://www.environmentaldefense.org/documents/523_starbucks.pdf (accessed March 2, 2005).

Steffan, James. 2003. Personal communication between James Steffen, chief engineer for vehicles, Federal Express, and the author. October 23.

Tomforde, A. 1994. Saxony's Green Fridge Frozen Out But Wine Goes Down Well. *The Guardian*, June 11, 39.

U.S. Congress, Office of Technology Assessment. 1994. Industry, Technology, and the Environment: Competitive Challenges and Business Opportunities, OTA- ITE-586. Washington, DC: U.S. Government Printing Office.

van der Linde, Claas. 1994. *Competitive Implications of Environmental Regulation in the Refrigerator Industry.* Washington, DC: World Resource Institute/Management Institute for Environment and Business.

Zuboff, Shoshana. 1984. *In the Age of the Smart Machine*. New York: Basic Books.

Part IV:
Conclusion

10

The Promise and Performance of Management-Based Strategies

Cary Coglianese and Jennifer Nash

In 1961, when President John F. Kennedy announced the goal of landing an astronaut on the moon by the end of the decade, the nation held great faith in modern technology's ability to achieve seemingly impossible objectives. Despite the occasional setback, that faith was rewarded with repeated missions landing astronauts on the moon. Yet, after decades of stunning, sustained successes in space exploration, the nation came to learn that technology by itself was, tragically, not enough to maintain progress in space. Presidential commissions investigating both the *Challenger* disaster in 1986 and the *Columbia* disaster in 2003 indicated that a safe and successful space program depends at least as much on effective management as on advanced space technology.[1]

The United States' approach to achieving ambitious environmental goals has traveled a path similar to that of its space program, although certainly one less dramatic. Federal environmental laws adopted during the 1970s placed a seemingly unbounded faith in technology to solve the nation's environmental problems. Expressing Kennedy-era optimism, Congress declared sweeping goals and imposed stringent standards intended to force the adoption of new technologies that would rescue the country from the grip of its environmental woes. As with the space program, policymakers' technological faith was rewarded with early successes. Yet, as with the space program, there has been growing awareness in recent years that technology by itself cannot be the answer to all environmental problems. People are ultimately the source of pollution, so solutions to environmental problems depend on shaping the incentives faced by the individuals who run and work in the private sector as well as on affecting the types of management within their organizations (Coglianese and Nash 2001; Potoski and Prakash 2005; Kagan 2006).

If future efforts to address environmental concerns are to aim increasingly inside the black box of the firm, then researchers and policy analysts will need to pay more attention to the kinds of management-based environmental strategies considered in this book. How should such strategies be designed? What is the role for government versus third parties in implementing management-based strategies? Should management-based strategies substitute for or just complement conventional forms of environmental regulation? Ultimately, the answers to these questions depend on the answer to a prior, still more fundamental question: Do management-based strategies work at all?

As Lori Snyder Bennear discusses in Chapter 3, there are compelling theoretical reasons to expect that management-based strategies can lead firms to improve their environmental performance. Management-based strategies call upon firms to invest in the production of information about the environmental risks created by their operations and about alternative mitigation measures, as well as to develop procedures for continued monitoring and information collection. The information generated through management systems leads to behavioral change either by (1) providing feedback directly to decisionmakers within firms about ways to reduce potential liabilities, or (2) giving information to government and other interested parties who in turn bring pressure to bear upon the firms' decisionmakers.

Management-based strategies seek to provide incentives for managers to invest in the collection of information. Without such incentives, firms may not always find it in their interest to gather the information they need to identify potential opportunities for environmental improvement, even when these opportunities might bring business advantages to the firm. This is because the search for such so-called win-win opportunities is not free. Even when the expected business benefits from such a search are positive, they may not be significant enough for managers to justify spending the time and resources needed to identify the win-win options in the first place. However, when government agencies or customers either mandate planning or offer firms incentives to engage in such planning, those firms will more readily invest in the search for potential win-win opportunities. Once firms undertake a search for such information, they will be inclined to implement those opportunities that they find benefit both the environment and the bottom line (Coglianese and Lazer 2003).

We begin this concluding chapter with a review of the evidence about what seems to be the most relevant policy question—namely, whether management-based strategies actually do work in practice. Although there are strong theoretical reasons for believing that management affects firms' environmental performance, more widespread use of management-based strategies should depend in the first instance on whether these strategies can be shown empirically to lead firms to improve their environmental performance. Fortunately, since many innovative management-based strategies have existed for at least a decade, they are ripe for the kind of careful empirical investigation exemplified by chapters in this book. After discussing the principal empirical findings on the impact of management-based strategies found in this book, we turn to the other salient issues and themes raised by the research collected in the preceding chapters, including how to design management-based strategies, the role for third parties in improving environmental management, and the choice between management-based and conventional environmental regulation.

The Impact of Management-Based Strategies

Management-based strategies, as we outline in Chapter 1, can be distinguished according to who uses them and whether certain management practices are mandated or just encouraged. For convenience, we labeled four distinct types of management-based environmental strategies: (1) regulation, (2) mandates, (3) incentives, and (4) pressures. The evidence compiled by various authors in this book has shown that some of these strategies can indeed yield improvements in industry's environmental performance. In particular, the strategies that require firms to engage in specific management practices, whether imposed by government or the private sector, appear to have the clearest and strongest effects.

During the early 1990s, 14 states implemented laws requiring industrial facilities to engage in pollution-prevention planning to reduce their use of toxic chemicals. Empirical evidence that Lori Snyer Bennear presents in Chapter 3 demonstrates that plants subject to such management-based regulations have experienced greater reductions in toxic chemical releases than they would have in the absence of these laws. Facilities subject to the pollution prevention planning laws reduced toxic chemicals by nearly 60,000 pounds more than comparable facilities not subject to management-based regulation.

Still further evidence indicates that requiring certain management practices can make a difference in shaping firms' environmental performance. Findings from a study of management-based mandates imposed by industrial customers on their suppliers, as Richard Andrews, Andrew Hutson, and Daniel Edwards, report in Chapter 5, suggest that private sector mandates can lead firms to improve their environmental performance in some, though not all, aspects of their operations. Drawing on a survey of 3,200 manufacturing facilities in four sectors, Andrews, Hutson, and Edwards compare the self-reported environmental performance of facilities that adopted formal EMSs in response to customer mandates with the self-reported performance of firms that were not subject to an EMS mandate and had not implemented one. Managers in companies that implemented an EMS in response to a business mandate reported greater improvements in energy use, recycling, and reductions in spills and leaks than managers in companies without an EMS.

Although requiring firms to adopt certain management practices can lead them to make some improvements, this does not mean that all management-based strategies will be equally effective. Nor does it mean that all effective strategies will lead to improvements across the board in firms' environmental impacts. For example, as the Andrews, Hutson, and Edwards study indicates, management-based mandates failed to result in any improvements in firms' self-reported performance with respect to air pollution, water pollution, or the production of hazardous wastes.

In addition, other studies reported in this book suggest that strategies that merely encourage firms to improve their environmental management will make much less of an impact than mandatory requirements. For example, Jason Scott Johnston reports in Chapter 7 the results from an empirical study of the effectiveness of EPA's Strategic Goals Program (SGP), a management-based incentive program established between EPA and the metal-finishing industry in 1998. SGP aimed to spur firms to adopt improved management practices and meet ambitious "beyond compliance" performance goals. Although upwards of 300 firms joined SGP and achieved reductions in environmental impacts compared with a 1992 baseline, Johnston concludes that the impact from the SGP program itself was at best limited. SGP did help to disseminate information about pollution prevention throughout the sector, but many of the environmental improvements made by participating firms occurred before the program was launched in 1998. Moreover, the program attracted only a small percentage of all the shops in the metal-finishing sector.

Another EPA management-based effort—the Clean Charles 2005 Initiative—owes its greatest impact to mandated requirements rather than to encouragement. As Tapas Ray and Kathleen Segerson describe in Chapter 8, EPA has viewed the Clean Charles 2005 Initiative as a successful project that has led to significant improvements in water quality of the Charles River in Boston. Yet EPA's efforts to focus attention on water quality actually were most effective by focusing the efforts of government on inspections and enforcement actions, not by giving dischargers' a clear management goal. The enforcement actions taken in response to the government's goal-setting, Ray and Segerson demonstrate, played the major role in spurring improvements in environmental management and water quality.

Overall, the studies in this book show that the impacts of management-based strategies can vary from clearly positive to marginal to nonexistent. A further challenge for both researchers and decisionmakers will be to understand why these strategies work in some circumstances but not others.

A key factor affecting the success of management-based strategies appears to be the adequacy of incentives. Mandates backed up by government or private sector sanctions appear to have much greater impact on firms' performance than strategies that merely encourage firms to improve their environmental management. Even so, there are undoubtedly limits to what can be achieved by the incentives accompanying management-based strategies. Mandating the adoption of a management system is not the same as mandating the adoption of new technologies or mandating the achievement of stringent performance standards that necessitate significant investments. In cases where costly changes are needed to improve environmental quality, it is not clear whether any management-based strategy can work, at least all on its own. In addition, many small firms simply lack the capacity to make any significant environmental improvements, regardless of even the strongest incentives available through management-based strategies. Large multinational firms often have the capacity to implement sophisticated and meaningful management systems, but similar progress probably cannot be expected of small- to medium-sized enterprises or businesses facing a high level of global competition or tight profit margins.

More attention to the costs of these management-based strategies is also needed, at least when management systems are mandated. The costs of complying with management-based mandates are presumably nontrivial. Some businesses have consumed tens of thousands of management hours complying with the paperwork requirements of the risk-management planning rules that Paul Kleindorfer analyzes in Chapter 4. Decisionmakers will need to determine whether the costs of generating information in response to management-based strategies are justified in light of the magnitude of environmental benefits that result. Ultimately, the net benefits of management-based strategies (that is, environmental benefits minus compliance costs) should also be compared with the net benefits from alternative environmental strategies, including more conventional forms of regulation.

The studies in this book indicate that management-based strategies can have an impact, but not in every situation where they are used. In those cases where management-based strategies have failed to result in observable changes in firms' environmental performance, researchers need to probe further to consider at least three alternative explanations for their null findings. First, the true impacts of management-based strategies may come from incremental across-the-board improvements in firms' environmental performance, rather than dramatic improvements in a single measure of their performance. Management-based strategies may result in smaller, though still positive, effects spread across a firm's entire operations. Researchers who choose a single measure of environmental performance may therefore fail to observe a statistically significant impact from management-based strategies, even though the cumulative effects of incremental improvements across all areas of operation, if they could be measured, might be substantial. In short, there may be positive effects from management-based strategies, but they may not be the ones that researchers have yet been able to observe and measure.

Second, the true impacts of management-based strategies, if any, may be observable only over a long time horizon. Even a decade or two might be too short to discern the impacts of improved environmental management. This is because the information generated through improved management takes time to flow throughout a firm and shape the thinking of its product designers and process engineers (Ehrenfeld 1998). The most substantial effects from management systems may not occur until the next time the company develops a new product line or production process.

Finally, even if some management-based strategies fail to yield any demonstrable results, this may well be because of the way these strategies have been designed. Firms may not have responded to the strategies because they provide insufficient incentives, or the strategies may have prompted firms to adopt certain management practices that are themselves ineffectual. The variation in results across the empirical studies in this book certainly suggests that the impact of management-based strategies probably depends greatly on the incentives they provide to firms and the way these strategies are designed.

The Design of Management-Based Strategies

Management-based strategies do not represent a single approach but are actually different kinds of efforts aimed at improving companies' environmental management and performance. Although they share the purpose of fostering effective management, these strategies can differ along at least two major dimensions, as we describe in Chapter 1. Whether the initiating institution is governmental or nongovernmental, and whether improved management is required or simply encouraged, can affect the ultimate impact of management-based strategies.

Choices in Designing Management-Based Strategies

Just as either of these two major dimensions could make a difference in the outcomes achieved, a variety of other potentially relevant differences in the design of management-based strategies may be important. These other design features include the following choices:

- *Planning With or Without Implementation.* Management-based strategies can encourage or require planning only (leaving it up to firms to decide on their own whether to implement some or all of their plans), or they can provide incentives for firms both to engage in planning and to implement their plans.
- *Types of Management Actions.* The types of actions required or encouraged by management-based strategies can vary. For example, some management-

based strategies call for employee training as part of the preferred management system, while others do not. Some call for managers to establish goals
consistent with clearly stated performance targets, while others do not stip-
ulate performance targets expected from managers and their goals.

- *Specificity of Actions.* Expectations for planning and management actions
  can be general or specific. For example, some pollution prevention planning
  laws call for firms to do little more than adopt "appropriate" plans, while
  other such laws call for firms to develop plans that meet detailed and extensive criteria.
- *Information Collection.* Different management-based strategies call for firms
  to collect different kinds of information. In addition, there are differences in
  whether information and records are to be kept by the firms themselves or
  should be released to others, including the public or government.
- *Auditing.* The extent of auditing varies, as does the type of auditor. Since the
  incentives offered by different management-based strategies are usually contingent on firms taking the specified management actions, some kind of
  auditing is needed. Verification can be conducted frequently or infrequently,
  on an announced or unannounced basis, and by government or third-party
  auditors.
- *"Stakeholder" Involvement.* For some management-based strategies, firms
  are expected to engage with community or environmental groups as part of
  their environmental management process. For other strategies, no community consultation is expected. The role of third parties may prove
  instrumental to the effectiveness of some strategies.

The key challenge in the development of management-based approaches to
environmental policy will be to identify which of these design elements, or
which combinations of these elements, yield the most successful outcomes
under specific conditions (Coglianese and Lazer 2003). In addition, when drawing inferences from empirical studies, the possible differences in the design of
management-based strategies should be taken into account. Even though a
research study may show that a particular strategy does (or does not) have an
observable impact on firms' environmental performance, it is possible that
other strategies with different design elements will yield different results.

Involvement of Third Parties in Environmental Management

Many observers believe that the involvement of community members, environmental groups, and other third parties is crucial to successful environmental
management (MSWG 2004). A larger role for external stakeholders in the
design and review of firms' environmental management systems may help
ensure that managers assess their firms' environmental impacts more fully, set
meaningful performance targets, devote adequate resources to system imple-

mentation, and establish mechanisms to identify and correct performance problems. Management-based strategies that attempt to institutionalize third-party involvement may also keep pressure on firms to make continuous improvements.

Why would firms even contemplate involving external stakeholders in their environmental management? For some firms, external involvement may help deliver important, although often intangible, reputational benefits. Involving outside groups in firms' internal management decisionmaking, however, presents numerous challenges. Determining which outsiders to include is not always easy. Deciding who can participate also means deciding who cannot participate, with the risk that excluded groups will feel worse than if no one had been allowed to participate at all. The question arises of whether firms need to develop procedures to provide public notice of changes to their management plans and allow outsiders opportunities to comment on those plans.

A further problem is that community and environmental organizations often lack the resources to make a meaningful contribution to companies' development and implementation of management systems. Local groups may lack sufficient technical expertise about facility operations, and the large, national environmental organizations that possess greater expertise lack the organizational presence and staffing needed to help design and monitor management at facilities across the country. Furthermore, the environmental group representatives seldom express much interest in management per se, tending instead to focus their efforts directly on facilities' environmental performance or the adoption of specific pollution-control technologies.

There is yet another problem with stakeholder involvement. The interests of environmental organizations and firms often are not directly aligned, as Andrew King argues in Chapter 9. As King emphasizes, environmental advocacy organizations have an interest in seeing innovative industrial practices adopted at one firm diffuse throughout an industry, while businesses' interests generally lie in keeping information about their operations from their competitors. However, this problem of misaligned interests may be reduced when environmental groups become involved in a firm's environmental management system, as opposed to the firm's technology or performance directly. Environmental management is usually not a firm's major source of competitive advantage, so the risks a firm faces from involving external stakeholders may not be as great as they would be for new technologies. Although the risks of engaging environmental groups in the development and certification of environmental management systems may be lower than they are for projects more central to the firm's core business, the benefits may also be greater. After all, what constitutes good management will often not be clear to an outside observer, such as a regulator or trade association, so the existence of stakeholder involvement in a firm's environmental management lends credibility to that firm's efforts.

A related issue in designing management-based strategies centers on the role of third parties in auditing firms' management practices. If firms are to be rewarded for adopting certain management practices, or punished for not doing so, it will be necessary to know which firms have acted in ways meriting reward or punishment. Involvement by community or environmental groups may lend some credibility to a firm's management efforts, but engaging their involvement is difficult and does not always equate to the same level of oversight that a comprehensive audit can provide.

Who should perform systematic audits of firms' environmental management? Government agencies have neither the resources nor the personnel for auditing management systems on a large scale. One potentially promising way to leverage private sector auditing capacities is offered by Howard Kunreuther, Shelley H. Metzenbaum, and Peter Schmeidler in Chapter 6 in the form of a proposal for a system of mandatory environmental insurance. Under such a regime, insurers would have incentives to conduct inspections of firms in order to make premium ratings and reduce claims. To a much greater extent than other third-party auditors, who are normally paid for their services by the audited firm, insurance companies have a financial interest in reducing risks. Furthermore, the costs of the inspections would be paid for in the premiums firms pay, thus providing a way to overcome the resource constraints that governments and community groups face. Although a mandatory insurance program would not necessarily substitute for conventional regulation and government enforcement efforts, it could complement them in a way that is likely to enhance the quality of firms' environmental management.

Management-Based Strategies and Conventional Regulation

A critical question surrounding management-based strategies is when to use them. Where do they fit into the existing array of policies and strategies for environmental protection? More specifically, what is their relationship to conventional regulation? The chapters in this book suggest that management-based strategies can be used both to help firms come into full compliance with existing regulations and to help them take steps that go beyond compliance with these regulations. In addition, management-based strategies may on occasion be appropriate alternatives to conventional regulation.

Management-based strategies can lead firms to improve their compliance with conventional regulations by encouraging them to identify the rules to which they are subject and to develop plans that will allow them to come into and maintain compliance. Such compliance management systems often include regular internal audits to identify and correct instances of noncompliance. Although some anecdotal evidence suggests that management systems do enhance firms' regulatory compliance, the study by Andrews, Hutson, and

Edwards in Chapter 5 finds no significant differences in the reported levels of compliance between firms with and without management systems. These results suggest that firms can and do come into compliance even when they do not have formal environmental management systems in place.

Perhaps management systems then should be viewed by companies as a means of reducing environmental impacts not currently addressed by government regulation. For example, the same study by Andrews, Hutson, and Edwards in Chapter 5 finds that the presence of management systems correlated with significant improvements in reported environmental impacts on unregulated aspects of business—such as spill avoidance or energy conservation—but did not correlate with reported improvements on regulated aspects, such as air and water emissions. These findings suggest the possibility that management-based strategies may be especially suited for environmental problems that call for improved operational management and internal coordination—problems that may be difficult to address through conventional regulatory strategies (Coglianese and Lazer 2003).

Management-based strategies implemented in the United States over the last decade have all been implemented against the backdrop of extensive government regulation. Management-based strategies may work only (or best) when they have a credible regulatory threat operating in the background. That threat can motivate firms to make progress on their own under the greater flexibility afforded by management-based strategies. For example, as Jason Scott Johnston indicates in Chapter 7, firms in the metal-finishing industry were motivated to participate in the Strategic Goals Program in order to preempt the adoption of tougher water pollution regulations that EPA had proposed in 1995. Government agencies can even explicitly use the existence of burdensome conventional regulation to offer rewards—namely, waivers from existing regulations—to firms that demonstrate responsible environmental management, something the EPA has attempted with its National Environmental Performance Track.

Although conventional environmental regulation has long been acknowledged to have many problems, the chapters in this volume raise possible problems that may afflict management-based strategies as well. Those outside of the firm may find it much too difficult to influence internal management and organizational behavior in any precise and meaningful way. Many management-based strategies call for systems and procedures that can be documented and reviewed by auditors. Yet good management involves much more than a flowchart or set of procedures that exist on paper. It reflects the dynamics of organizations made up of people and their relationships with each other (Howard-Grenville 2005). Even the most informed government officials will never be as well-situated as private sector managers to know the best way to manage businesses so as both to return a profit and to minimize impact on the environment. Moreover, government and outside groups are far from unified, so the possibility exists that different agencies and organizations could require duplicate or incompatible management steps.

Furthermore, some of the problems with management-based strategies could turn out to be worse than some of the problems associated with conventional regulation. Strategies that take the form of government requirements for certain management processes intrude into the core of business decision-making. Putting management into private hands is, after all, what a free enterprise system is all about. Government-imposed standards on environmental management could turn out to be too rigid, especially in the face of changing conditions in global markets. Although there is much to be said for the kind of systematic, centralized, and rational management practices called for by many current management-based strategies, it is entirely possible that the best management practices—whether these practices are for preventing accidents, reducing pollution, or expanding shareholder value—require a much greater degree of decentralization and internal competition. If nothing else, those who seek to use management-based strategies to improve environmental performance need to be mindful of the possible negative effects these strategies could have not only on firms' overall ability to expand productivity and shareholder value, but even on their ability to achieve environmental protection goals.

The potential shortcomings and costs of management-based strategies are certainly not trivial, so decisionmakers need to assess whether the advantages of these strategies outweigh their disadvantages. Even though several empirical studies show that management-based strategies can lead to environmental improvements (Andrews et al. 2006; Bennear 2006; Ray and Segerson 2006), the substantive significance of these improvements still needs to be assessed and compared with other alternatives. Whether the benefits achieved through these strategies are worth the costs that they impose on economic activity remains an important subject for future research.

In making decisions about management-based strategies, decisionmakers should take into account the full range of possible impacts these strategies may generate, both positive and negative. An unintended consequence of trade association mandates, for example, could be that numerous marginal firms leave the trade associations that have imposed such mandates, perhaps making such firms less visible to regulators, competitors, and others who could monitor their environmental impacts. A similar effect might also arise with government programs when firms take actions to bring their use of specified chemicals below levels that trigger the imposition of management-based regulation— even if doing so does not substantially lower their overall level of environmental risk (Bennear 2005). Firms that fall below triggering levels for government reporting schemes may also tend to fall off the radar screens of regulators and community groups. To address these kinds of side effects, management-based strategies should probably be combined with other efforts by government, trade associations, and community groups to keep firms from shifting their operations or evading monitoring.

A final question that can be raised about management-based strategies focuses on equity—both from the standpoint of the public as well as of industry itself. For the public, the flexibility inherent in a management-based approach may mean that the same types of facilities could emit different levels of pollution in different locales. Although conventional regulation may be criticized for taking a one-size-fits-all approach, uniform technology- or performance-based standards are at least uniform. For business, there is a separate equity concern: government inspectors and others who oversee management-based programs will apply management standards in inconsistent or inequitable ways. If what counts as good management is not clearly specified, this will give discretion to auditors and may result in an uneven application of sanctions or rewards (Coglianese and Lazer 2003).

Equity issues also arise from the distinction we address in Chapter 1 between actions and attitudes. Management-based strategies work by discriminating among firms based on whether they have in place certain easily observable management practices. Consequently these strategies will be vulnerable to the criticism that firms are selected for reward or punishment based on the wrong criteria. Some firms may be rewarded simply because they go through the motions of adopting a management system, while other firms that are really making a difference in reducing pollution could go unrewarded because they lack the requisite formalities in their management practices.

Conclusion

As we have seen, management-based strategies are increasingly gaining the attention of leaders in both the public and private sectors. These strategies can take many forms, but they are linked by their emphasis on improving management *per se* and thereby seeking to contribute indirectly to improved environmental outcomes. These strategies hold out the promise that firms can be encouraged to gather information needed to improve their environmental performance—and that they will respond to the acquisition of new information by reducing or preventing pollution. By providing incentives for firms to identify their own risks and select their own mitigation solutions, management-based strategies are flexible and seek to use the private sector's informational advantage for the public good.

The research this book presents suggests that management-based strategies, although still relatively new and overlooked, do have a role to play in environmental protection. Management style can shape the environmental performance of firms (Kagan 2006), so strategies that influence private sector management are plausible candidates for bringing about environmental improvements. The studies here break new ground by empirically investigating the impact of management-based strategies. Some, but not all, of these studies confirm that management-based strategies can contribute to reductions in the environmental impacts of firms (Andrews et al. 2006; Bennear 2006).

Of course, the overall impact of any strategy depends on a variety of factors, including the incentives it provides to firms to make improvements and the type of environmental problem being addressed. Moreover, it is clear that management-based strategies have their disadvantages as well as their advantages. As such, policy analysts should be cautious about overstating what can be accomplished through management-based strategies. Radically altering the existing system of environmental regulation in favor of adopting only management-based strategies would certainly be ill-conceived. Rather, the challenge for decisionmakers will be to find the optimal intervention for the specific problems and circumstances they confront. In some cases, the best option will be to continue to rely on conventional regulatory strategies. And if there are problems with conventional regulation, the proper solution often will be to fix those problems directly—not to expect that management-based strategies will make up for all the shortcomings of the existing regulatory system. After all, an excessive emphasis on management-based strategies could itself result in pre-empting other regulatory interventions that would better serve society.

Yet in some cases, as chapters in this book suggest, there will be good reason to consider using a management-based strategy. Management-based strategies may be particularly useful when policymakers seek to influence the practices of a highly diverse set of facilities (Coglianese and Lazer 2003), collect information that will help motivate private sector managers or activate influential stakeholders (Bennear 2006), or improve performance among facilities or with respect to specific problems that are simply not amenable to other regulatory approaches (Andrews et al. 2006).

Management-based strategies have been shown to yield improvements in industry's environmental performance in certain cases, indicating that anyone concerned about environmental quality should seriously consider the use of these strategies. Surely the goal of environmental protection is too important to ignore tools that have been shown to make a difference. In the same way that NASA's accidents over the past two decades have tragically revealed the importance of institutional factors, such as internal management, to the success of the space program (Vaughan 1997), it seems clear than relying solely on technology to "fix" our environmental problems will not be enough. If we are to address successfully both persistent and emerging environmental problems, we must give full consideration to new strategies and apply them whenever appropriate. The challenge for researchers lies in figuring out just when that is.

Notes

1. NASA's management problems are addressed in Columbia Accident Investigation Board Report (2003) and Report to the President by the Presidential Commission on the Space Shuttle Challenger Accident (1986).

References

Andrews, Richard N. L., Andrew Hutson, and Daniel Edwards, Jr. 2006. Environmental Management Under Pressure: How Do Mandates Affect Performance? In *Leveraging the Private Sector: Management-Based Strategies for Improving Environmental Performance*, edited by Cary Coglianese and Jennifer Nash. Washington, DC: Resources for the Future, Chapter 5.

Bennear, Lori Snyder. 2005. Strategic Response to Regulatory Thresholds: Evidence from the Massachusetts Toxics Use Reduction Act (June 27, 2005). http://ssrn.com/abstract=776504 (accessed November 18, 2005).

Bennear, Lori Snyder. 2006. Evaluating Management-Based Regulation: A Valuable Tool in the Regulatory Toolbox? In *Leveraging the Private Sector: Management-Based Strategies for Improving Environmental Performance*, edited by Cary Coglianese and Jennifer Nash. Washington, DC: Resources for the Future, Chapter 3.

Coglianese, Cary, and David Lazer. 2003. Management Based Regulation: Prescribing Private Management to Achieve Public Goals, *Law and Society Review* 37 (4): 691–730.

Coglianese, Cary, and Jennifer Nash, eds. 2001. *Regulating from the Inside: Can Environmental Management Systems Achieve Policy Goals?* Washington, DC: Resources for the Future.

Ehrenfeld, John. 1998. Cultural Structure and the Challenge of Sustainability. In *Better Environmental Decisions: Strategies for Governments, Businesses, and Communities*, edited by Ken Sexton, Alfred A. Marcus, K. William Easter, and Timothy D. Burkhardt. Washington, DC: Island Press, 223–44.

Howard-Grenville, Jennifer A. 2005. Explaining Shades of Green: Why Do Companies Act Differently on Similar Environmental Issues? *Law & Social Inquiry* 30(3): 551–81.

Kagan, Robert A. 2006. Environmental Management Style and Corporate Environmental Performance. In *Leveraging the Private Sector: Management-Based Strategies for Improving Environmental Performance*, edited by Cary Coglianese and Jennifer Nash. Washington, DC: Resources for the Future, Chapter 2.

MSWG (Multi-State Working Group on Environmental Performance). 2004. The External Value Environmental Management System Voluntary Guidance: Gaining Value by Addressing Stakeholder Needs. http://www.mswg.org/documents/guidance04.pdf (accessed August 9, 2005).

NASA (National Aeronautics and Space Administration). 1986. Report to the

President by the Presidential Commission on the Space Shuttle Challenger Accident. http://history.nasa.gov/rogersrep/genindex.htm (accessed November 18, 2005).

———. 2003. Columbia Accident Investigation Board Report. http://www.nasa.gov/columbia/home/index.html (accessed November 18, 2005).

Potoski, Matthew, and Aseem Prakash. 2005. Covenants with Weak Swords: ISO 14001 and Firms' Environmental Performance. *Journal of Policy Analysis and Management* 24(4): 745–69.

Ray, Tapas K., and Kathleen Segerson. 2006. Clean Charles 2005 Initiative: Why the "Success"? In *Leveraging the Private Sector: Management-Based Strategies for Improving Environmental Performance*, edited by Cary Coglianese and Jennifer Nash. Washington, DC: Resources for the Future, Chapter 8.

Vaughan, Diane. 1997. *The Challenger Launch Decision: Risky Technology, Culture, and Deviance at NASA*. Chicago: University of Chicago Press.

Index

pollution-prevention planning of, 55–59

policy toolbox, 7–9

pollution-prevention planning, 55–59

Portland Cement Association (PCA), 11–12, 13

POTWs (publicly owned treatment works), 168–169, 173–174, 179, 191–192, 198n

precursors, 159
see also insurance

preemptive regulatory initiatives, 61–62

principal-agent model, 59–60

private inspections. *see* inspection programs

property transfer liability insurance, 145–146
see also insurance

publicly owned treatment works (POTW), 168–169, 173–174, 179, 191–192, 198n

pulp mills, 34–43, *39, 40, 41*

raw materials, 115–117

RCRA (Resource Conservation and Recovery Act) (1976), 176–177

reflexive law, 8–9

Regulating from the Inside (Coglianese and Nash 2001), 15

regulations
and insurance, 148–150
"license to operate" model, 33–34, 39
and management-based strategies, 19, 257–260
metal-finishing industry, 172–178, 177–178
pulp mills, 34–43, *39, 40, 41*

relative performance standards, 85n

Resource Conservation and Recovery Act (RCRA) (1976), 176–177

Responsible Care initiative, 4, 13, 15, 105

Right-to-Know Act facilities, 94, 95

Risk Management Programs (RMP) rule
accident epidemiology, 89
background, 88–92
and facility characteristics, 92–98
risks, 100
see also management-based regulations (MBRs)

risk-reducing management, 143–148, 237–241, *241*
see also insurance

risks
and MBRs, 73–74, *74*
uncertainty, 76–77
see also insurance

SC Johnson, 235–236, 237–241, *241*

self-regulation. *see* voluntary programs

social licenses, 33, 39, 41, 91–92

socioeconomic status (SES), 99
see also host communities

specification standards, 7

stakeholder involvement, 255–257

Starbucks, 236, 237–241, *241*

stormwater, 204, 206

Strategic Goals Program (SGP), 168–169, 178–194, *183*

strategic regulatory preemptions, 185

supply chain mandates, 115–117, 185
see also mandates

synergies of excellence, 104

technology, promise of, 249

technology standards, 7

Terrorism Risk Insurance Act (TRIA) (2002), 150

third party involvement, 255–257

toxic chemicals, regulation of, 52, 63–64, 65, 66, 78–79

Toxic Use Reduction Act (TURA) (1989) (MA), 9–10, 62

Toxics Release Inventory (TRI)
chemicals reportable under, 86n
compliance, 32, 78
and evaluation of MBRs, 63
and mandates study, 120
metal-finishing industry, 177

trade associations, 11–12, 186–187
see also voluntary programs

triple bottom line, 113

underground storage tank coverage, 143

voluntary insurance. *see* insurance

voluntary programs
American Chemistry Council (ACC), 13, 15